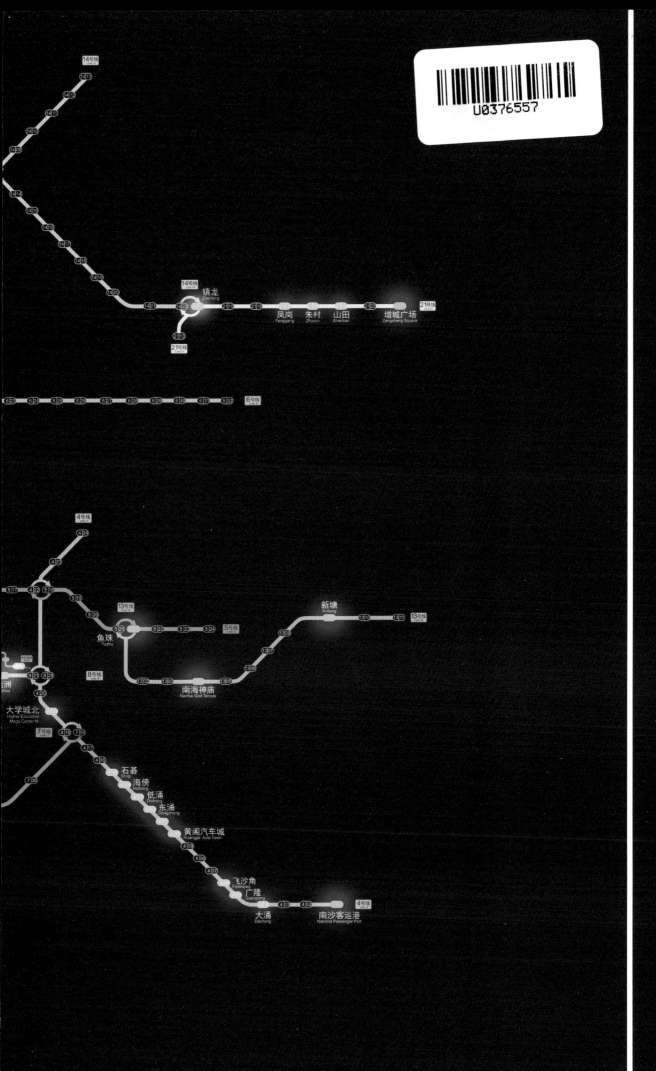

缘起广州，深耕二十五载，匠心设计，助力轨道未来

Deep cultivation originated in Guangzhou for 25 years, building the future of rail transit with ingenious design

求真、尽善、致美，谨以此书致敬新中国成立 70 周年

In pursuit for the true, the good and the beautiful, this book is dedicated to the 70th anniversary of the founding of the People's Republic of China

交通建筑作品
ARCHITECTURAL METROS

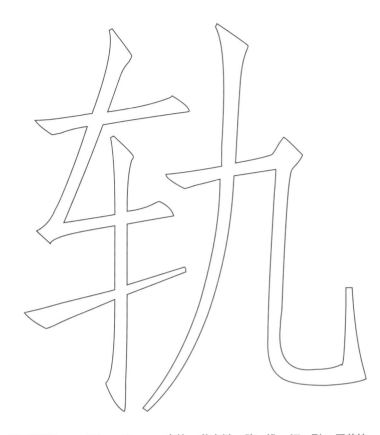

主编 曾宪川 陈雄 江刚 罗若铭
Chief Editors Zeng Xianchuan Chen Xiong Jiang Gang Luo Ruoming

广东省建筑设计研究院
Architectural Design and Research Institute of Guangdong Province

中国建筑工业出版社
CHINA ARCHITECTURE & BUILDING PRESS

序言 Foreword

《轨·道 交通建筑作品》是一部城市轨道交通建筑设计的经典之作，作者提出了对城市轨道交通建筑设计的理解、探索和实践，对广东省建筑设计研究院参与城市轨道交通建筑设计25年的设计方案和成功作品进行了分析总结，从一个侧面反映了我国城市轨道交通建筑设计的成长历程和成就，值得业内同行参考借鉴。

凭借在建筑设计领域的经验、成就和积淀，广东省建筑设计研究院经过25年的潜心研究、勇于创新和工程实践，针对城市轨道交通不同类型的建筑工程，将运营功能、交通需求和环境景观要求较好地体现在建筑设计作品中，为广州地铁和其他城市轨道交通建设创作了不少精品工程。

这部著作图文并茂，按"持续之道"、"兼容之道"、"文化之道"、"多元之道"、"创新之道"，对每项工程设计作品如何将建筑与功能、环境、文化、艺术和结构等完美地结合予以深入剖析，系统地总结了该院25年来城市轨道交通不同建筑类型的设计经验，为我国城市轨道交通工程设计领域的佳著之一，充分展现了该院在城市轨道交通建筑设计领域的技术实力和"求真、尽善、致美"的院风。

中铁二院轨道交通资深专家：孔繁达　周勇

2019年8月

Architectural Metros is classic in architectural design of urban transit. The author proposes the understanding, exploration and practice on architectural design of urban rail transit, analyzes and summarizes the design and successful projects of urban rail transit done by Architectural Design and Research Institute of Guangdong Province in 25 years, reflecting the development process and achievement of China's architectural design of urban rail transit, which gives a good reference to its peers.

With profuse experience, achievement and accumulation in the field of architectural design, through 25 years' diligent study, bold innovation and steadfast practice, Architectural Design and Research Institute of Guangdong Province has properly integrated operation functions, traffic requirements and environmental landscapes into architectural design for various architectural projects in different types of urban rail transit, and created a number of excellent works for Guangzhou Metro and other urban rail transit constructions.

With excellent pictures and texts, Architectural Metro has made thorough analysis on how to perfectly combine architecture with function, environment, culture, art and structure etc. for each project design in accordance with "the way of continuity", "the way of compatibility", "the way of culture", "the way of diversification" and "the way of innovation", and systematically summarized the design experience of various architecture types in urban rail transit by the Institute for 25 years. As one of the best works of China's urban rail transit design, it fully demonstrates the technical strength of Architectural Design and Research Institute of Guangdong Province in the field of architectural design of urban rail transit, and the institute spirit of "Pursuing the true, the good and the beautiful".

Kong Fanda, Zhou Yong Senior Expert in Rail Transit of China Railway Eryuan Engineering Group Co.,LTD.

August, 2019

本人与广东省建筑设计研究院的合作始于1994年广州地铁一号线广州东站，2012年又在港珠澳大桥工程中共同奋斗，今天见证他们历时25年的轨道交通设计成果结晶——《轨·道 交通建筑作品》，彼此可谓结下了深厚缘分。

25年来，广东省建筑设计研究院始终秉承敢试敢闯、敢为人先的创新精神和"求真、尽善、致美"的设计理念，为广州地铁设计了动物园站、琶洲站（装修）、南沙客运港站（装修）等一批经典车站，成为广州轨道交通设计不可替代的设计合作单位之一。

《轨·道 交通建筑作品》一书展示了25年来广东省建筑设计研究院的轨道交通设计历程，提炼了轨道交通设计之"道"，将轨道交通设计从交通工程设计上升为文化及艺术的建筑设计，体现了广东建筑设计研究院深厚的人文积淀和探索传统，也契合了新时代公众对轨道交通的美好向往和追求。

祝愿广东省建筑设计研究院在未来取得更加令人瞩目的成就。

广州地铁原总工程师：陈韶章

2019年8月

I have been working with Architectural Design and Research Institute of Guangdong Province since 1994 on Guangzhou East Railway Station of Guangzhou Metro Line 1, then we again collaborated on the project of Hong Kong–Zhuhai–Macao Bridge in 2012, and now, I witness the publication of *Architectural Metros*, the great achievement of rail transit design after their 25 years' hard work. There is a deep connection between us.

For 25 years, Architectural Design and Research Institute of Guangdong Province has been sticking to the aggressive innovation of making attempts and pioneering sprit, and the design philosophy of pursuing the true, the good and the beautiful, therefore created a series of classic stations such as the Zoo Station, Pazhou Station (decoration), Nansha Passenger Port Station (decoration) and so on, and become one of the irreplaceable design cooperation organizations of Guangzhou metro design.

Architectural Metros demonstrates the process of metro design by Architectural Design and Research Institute of Guangdong Province in 25 years and extracts the "ways" of metro design, upgrades metro design from traffic engineering design to architectural design of culture and art. It also reflects the in-depth humanities accumulation and exploration tradition of Architectural Design and Research Institute of Guangdong Province and meets the public's aspiration and pursuit for the better metros in the new era.

Here I wish more remarkable and splendid achievement for Architectural Design and Research Institute of Guangdong Province in the future.

Chen Shaozhang, Former Chief Engineer of Guangzhou Metro
August, 2019

前言

1994年，广东省建筑设计研究院开始广州市轨道交通车站站点设计；
1997年6月28日，广州市轨道交通一号线广州东站开通运营；
2002年12月28日，广州市轨道交通四号线大学城北站开通运营；
2005年12月28日，广州市轨道交通大学城专线北亭站开通运营；
2006年6月28日，广州市轨道交通三号线广州塔站开通运营；
2006年12月28日，广州市轨道交通三号线汉溪长隆站开通运营；
2006年12月28日，广州市第一条轨道交通高架线路——广州市轨道交通四号线石碁站、海傍站、东涌站、低涌站、黄阁汽车城站开通运营；
2009年12月28日，广州市轨道交通五号线坦尾站、动物园站开通运营；
2010年12月28日，广州市轨道交通二、八号拆解工程线黄边站、萧岗站、飞翔公园站、白云文化广场站、江夏站、白云公园站开通运营，广州市轨道交通三号线机场南站开通运营；
2017年12月28日，广州市轨道交通四号线南延段飞沙角站、广隆站、大涌站开通运营；广州市轨道交通九号线飞鹅岭站、花都汽车城站开通运营；
2018年6月28日，广州市轨道交通三号线机场北站开通运营；
2018年12月28日，广州市轨道交通二十一号线山田站、朱村站开通运营。

二十五载，艰辛探索，玉汝于成。

2000年12月，广州市轨道交通二号线广州火车站结构封顶，创造了中国地铁建设史上8个月车站主体封顶的建设记录；
2002年，广州市轨道交通二号线开通在即，业主紧急请求技术援助，我院义无反顾，派出技术人员全力以赴力保开通；广州第一条高架线——广州市轨道交通四号线，首期五站由我院全程主导；
2009年，广州市轨道交通五号线动物园站，以中国首创第一例中庭式站厅、叠层式车站、清水混凝土装修惊艳亮相；广州市轨道交通四号线坦尾站又以中国第一例高架与地下紧密换乘车站横空出世；广州市轨道交通五号线区庄站、西村站、西场站，地面建筑方案多次审查未能通过，我院再次应业主要求临危受命，最终确保开通，不辱使命；

2017年，广州市轨道交通十三号线车站,装修设计做了大胆突破——南沙客运港站的"宝船起航"和南海神庙站的"海不扬波"，两个文化车站场景，带给地铁车站全新体验。

二十五载，形于轨，成于道。

坚持，坚守；原创，人文；
地铁改变了城市格局，提升了城市品质；
地铁设计也改变了广东省建筑设计研究院；
开拓，提升；进取，加速；
广东省建筑设计研究院与广州地铁已结下深厚的缘分
从普通民用建筑到轨道交通行业，从新手到行家；
遵循严格的技术要求与逻辑，为市民提供安全、便捷、整洁的交通场所；
以人为本，文化融入，不断创新，为乘客提供富有文化内涵的精神体验场所；
形态可模仿，精神方永恒；
源于技，不拘于技，不断创新。

二十五载，修桥筑路，福荫后人。

潜心钻研轨道设计之道，同心协力解决技术难题；
设计成果反复修改与比选，严格论证，务求精益求精；
设计任务复杂，多部门跨专业配合协作，经重重审查锤炼成果；
一线工地现场紧密配合，挥洒努力与理想的汗水；
默默耕耘，为地铁设计事业青春无悔。

二十五载，不忘初心，继续前行。

因为自信从不犹豫；
因为坚持从不徘徊；
因为热血从不停步；
因为理想从不妥协；
因为使命从不后退。

Preface

In 1994, Architectural Design and Research Institute of Guangdong Province began to design the stations of Guangzhou Metro System;
June 28, 1997, Guangzhou East Railway Station of Guangzhou Metro Line 1 opened for operation;
December 28, 2002, Guangzhou Railway Station of Guangzhou Metro Line 2 opened for operation;
December 28, 2005, Higher Education Mega Center North Station of Guangzhou Metro Line 4 opened for operation;
June 28, 2006, Canton Tower Station of Guangzhou Metro Line 3 opened for operation;
December 28, 2006, Hanxi Changlong Station of Guangzhou Metro Line 3 opened for operation;
December 28, 2006, Guangzhou's first elevated metro line — Shiqi, Haibang, Dongchong, Dichong and Huangge Auto Town Stations of Guangzhou Metro Line 4 opened for operation;
December 28, 2009, Tanwei and Zoo Stations of Guangzhou Metro Line 5 opened for operation;
December 28, 2010, Huangbian, Xiaogang, Feixiang Park, Baiyun Culture Square, Jiangxia and Baiyun Park Stations of Separation Engineering Line of Guangzhou Metro Line 2 & Line 8 opened for operation; Airport South Station of Line 3 opened for operation;
December 28, 2017, Feishajiao, Guanglong and Dachong Stations of Guangzhou Metro Line 4 South Extension opened for operation;Fei'eling and Huadu Autocity Stations of Guangzhou Metro Line 9 opened for operation;
June 28, 2018, Airport North Station of Guangzhou Metro Line 3 opened for operation;
December 28, 2018, Shantian and Zhucun Stations of Guangzhou Metro Line 21 opened for operation.

For 25 years, the arduous explorations have led to abundant accomplishment
In December 2000, the structure of Guangzhou Railway Station of Guangzhou Metro Line 2 sealed the top, creating the record in China's metro construction history of sealing the top of a station's main structure within 8 months;In 2002, when Guangzhou Metro Line 2 was going to open soon, we were urgently requested for technical assistance. We did not hesitate to send our technical team and buckle down to ensure the timely opening of Line 2. The five stations of Guangzhou Metro Line 4 Phase 1, the first elevated line in Guangzhou was led by Architectural Design and Research Institute of Guangdong Province. In 2009, Zoo Station of Guangzhou Metro Line 5 was unveiled being the first atrium–style station hall originated in China with terraced station and as–cast–finish concrete decoration. Tanwei Station of Guangzhou Metro Line 4 then debuted as the first interchange station seamlessly connecting elevated and underground tracks in China the ground construction plan of Ouzhuang, Xicun and Xichang Stations of Guangzhou Metro Line 5 had failed to pass the censor several times, we accepted the task at this critical moment and successfully ensured the opening in the end.

In 2017, Guangzhou Metro Line 13 opened with bold and break through decoration design — "the treasure ship setting sail" of Nansha Passenger Port Station and "the calm and tranquil sea" of Nanhai God Temple Station. The two cultural station settings brought brand new experience to the metro stations.

For 25 years, the track has been constructed, and the metro system has been built
Perseverance, persistence; originality, humanities;
Metro has changed the structure of the city, and enhanced the quality of the city;
The metro design has also changed Architectural Design and Research Institute of Guangdong Province;
Development, improvement; aggressiveness, acceleration;
There has been strong connection grown between Architectural Design and Research Institute of Guangdong Province and Guangzhou Metro;
From ordinary civil architecture to the rail transit industry, from novice to expert;
We not only follow strict technical requirements and logic to provide safe, convenient and clean transportation for citizens;
But also stick to people–oriented, culture–based and innovative practice to provide public space of cultural connotation and spiritual experience to passengers;
The form can be imitated, but the spirit is eternal;
The design is derived from but not constrained by technology.
Continuous innovation is the only and final answer.

For 25 years,we have built bridges and roads to benefit the future generations
We dived into metro design and worked together to solve technical problems;
The design results were repeatedly revised, selected and strictly demonstrated to make perfection more perfect;
The design tasks were complex demanding multi–sectoral interdisciplinary collaboration, before being recognized and approved through multiple censors;
The workers on the construction sites closely coordinated and made great efforts to achieve our common goals;
We dedicated to metro design cause through quiet diligence with no regrets.

For 25 years,we stayed true to our mission and kept on moving
Because——
Confidence never hesitates;
Persistence never halts;
The blood never stops;
The ideal never compromises;
The mission never retreats.

持续之道
（地下车站）

1. 持续的空间

城市地铁的空间形态，最初是以单个站点形式进行建设，站点通过隧道连成线路，各条线路通过换乘站点构成轨道交通的网络。空间形态呈现"点""线""面""体""网"的持续生长趋势。

"点"的形态

车站的地面附属设施如风亭、出入口、冷却塔、小型地面站厅（或者与部分服务设施结合）等，是车站功能的聚集处，也是功能流线及客流流线的末端或节点。在城市空间格局中呈现"小""散""多"的特色。

"线"的形态

车站的隧道、暗挖站台、人行通道、天桥等是车辆或客流穿行的空间，具有高度流动性及线性化的特征。这些"线"的存在把貌似分散的"点"或隐（地下）或现（地面或高架）或半隐半现地"连接"起来。

"面"的形态

多线换乘的车站，车站站厅规模较大，虽然车站空间高度的绝对值相比普通车站高，但高度与面积的比例远比一般车站小，呈现出扁平化的空间形态，成为平面型空间。

大型的车辆段（基地）占地规模较大，城市空间形态也呈现高度扁平化的特征。后期再在上方作上盖的开发，整个车辆段（基地）则成为基座。

"体"的形态

高架车站（不论在路中还是路侧的车站），车站站台加上车站的顶盖都会形成较为明显的体量感，成为城市景观的视觉焦点，某些整体式设置（路中设备房、站厅、站台合为一体）体量感更加明显。

某些地下车站，因外界条件影响，在地下部分设置了高大的中庭式地下空间，从而在地下形成了巨大的"体块"。

"网"的形态

区域整体线路的规划及建设，使地铁车站从线路的延伸及扩张，逐渐形成覆盖区域性的线网。

这也是有别于大多数公共建筑及交通建筑的（如机场、码头、公交车站等），也仅有轨道交通车站能实现这种持续的生长，最终成为网状的存在。

这种网的形态，既可以说是完成的，也可以说是未完成的。线路一次建成，可以说是完成的形态。线路不是一次完成（广州市轨道交通二、八号拆解工程线）甚至还要经过拆解才能形成最终的形态。这是一种动态变化、自我调整的过程。

如果把所有地铁网络连通的部分（地上及地下，隧道及车站）看成一个整体的话，这个建筑的体量和规模无疑将是这个城市中最大和覆盖面最广的。它几乎涉及一个城市大部分区域，并且还在不断地变化扩张，成为一个"超级建筑"；如果单从地面呈现的分散出入口和配套设施来看，一般人是无法想象上述情形的。另一方面，这个"超级建筑"的表象，在局部却是极为"微不足道"的，有时仅仅是一个几平方米的地面风亭。

客观条件催生车站空间

地下车站与普通民用建筑物最大的区别在于无法在一块独立完整的地块中进行计与施工。车站的选址往往处于车流繁忙的城市干道地下区域，地下城市管网布。方案上也需要提前充分考虑，尽可能减少拆迁及施工对城市交通及周边环的巨大影响，这对设计师所掌握的技术的全面性及其宏观判断能力是一个非常巨大的挑战。同时，车站的选址需要对周边客流使用效率和商业价值进行分析估，通过多方案比选而得出最优结果。在车站设计上，可充分考虑施工条件素，将不利因素变为有利因素，使车站空间得到独特的表达。如广州地铁五号动物公园站，由于车站明挖且埋深较大，车站站厅局部公共区有条件形成高空间，结合空间中两个"Y"字形结构柱，为动物园站打造出个性鲜明的车站间。

2. 持续的时间

与其他类型的公共建筑相比，地铁车站有明显的流动性、过程性的动态特征（他公共建筑更多的是驻留式的静态功能性空间，如医院的诊室、病房，学校的室，超高层的办公室，酒店的餐厅、客房，文化中心的剧场等。公众在上述场有确定的目的性，相对都会停留一段较长的时间，场所特征呈末端化特性）。对于其他的公共建筑，地铁车站也有较强的目的性，乘客往往会选择前往某个站或在某个车站换乘，而到达后又会迅速离开，中间停留时间一般在几分钟内，呈现一种高度流动性的特征。因此场所特征是过程化的、持续的。

3. 持续的体验

与地面建筑相比，车站空间的使用者多数是从"内"着眼。其空间的单向性，乘客无法感受到室外的场景及时间变化；和其他公共建筑的"立体"体验相比，地铁车站的空间场所大多是扁平化和界面感的，场所体验有较强的平面化和场化特征。这种场景化常常可以与车站所在的地域相背离，形成自我一体的场所达。大多数车站公共空间具有相似的特征，乘客在穿越若干地下车站的站点时，无法感受地面或外界环境的变化。由整个城市的轨道交通线网来看，地下车站区间所形成的是一个连续而整体的空间，乘客在区间与车站中不断切换。容易成"出入口"—"通道"—"站厅"—"站台"—"车厢"—"隧道"—"台"—"站厅"—"通道"—"出入口"的持续的模式化体验。但由于不同车的尺度与空间形式均存在匀质化特征，乘客在地下轨道交通通勤过程中，其空间体验大致相同，且缺乏对周边环境的感知，易形成沉闷的感受。由人的体验角来说，车站空间可根据自身条件，营造出一定的差异性。一方面，车站的空间式结合环境、地质与结构等不同限制，打造出车站一定的差异性。另一方面，站装修结合地域差异，表达一定的文化意象，以为乘客提供丰富的体验为目标，寻求车站在空间、形式、尺度等方面的差异性，避免千篇一律标准站带来的平的乘车环境。

持续的生长

2005年,《广州市轨道交通建设规划》提出,线网规划15条线路,619km;2020年,《广州市轨道交通建设规划》提出,线网规划19条线路,817km;地铁建设的目标定位从服务城市的辅助基础设施逐渐向引领城市发展的先行,从单线建设到多线同时实施再到全网络的建设拓展,从单一地铁站场建设到上下联动综合交通枢纽及上盖物业,建设规模和涉及的范围区域也在不断扩大调整。

密集的地下城市轨道交通车站为这些庞大的地下开发提供了积极的交通条件及庞大的人流,依托地下车站,往往可以形成具有活力的地下商业街道,与地上商业形成联动发展。与此同时,紧凑的城市发展模式也给车站提出了新的要求与限制条件,对车站适应性的要求也越加强烈。

早期地铁建设模式是与城市建设并行的大拆大建,是以地铁建设空间需求为主的单向扩展,因此早期地铁车站形态更多的是两层标准车站。而随着城市的高速建设与扩展,地面建设的规划实施速度也远远高于地下空间建设的规划,导致地下车站的建设受到地上空间形态的约束,同时由于城市交通强度的变化带来对施工用地的日趋紧张的限制,车站形态也越来越呈现个性化为主、标准化为辅的非标准化趋势。

在广州城市地下深处,地铁线路已经像老树盘根一样覆盖着城市的大部分重要区域,而地下的每一个车站每天都承担着城市最大的交通运力和客流吞吐,已成为城市中市民参与度最高的建筑物。地下地铁车站不同于一般的地上交通建筑,地铁车站不存在外立面与造型,无法给城市带来标志性或直接的视觉冲击感,但它承载的功能及对城市的意义是其他建筑无法比拟的。如果说城市的博物馆、图书馆、艺术馆属于城市精神及意识形态的上层建筑,那么地铁车站就是支撑起整个城市上层建筑的地下基础。每一个车站就像城市地下的一个低调而高效运转的精密机器零部件,然后通过整体线网将所有零件连成一体,犹如一个埋于城市地中的巨型发动机。发动机运转也带动着城市的运作,直接影响着市民的活动出行、工作就业、居住饮食、商业发展、金融经济及城市的扩张与延伸。

车站作为城市的脉络,宏观上联系着城市的不同区域,乃至不同的城市,微观上也与周边环境紧密联系。能否与周边区域环境密切联系并产生互动,是评价一个车站效率与价值高低的重要因素。因此如何与周边区域保持持续发展的联系,适应车站周边发展带来的变化,是车站设计需要考虑的重要方面。车站与周边,除了车站通道等物理空间上的联系,还应当注重车站改变的可能性,使车站在日后发展过程中,能够有条件改变并适应周边持续开发。除此之外,客流的预判也是车站能否适应周边发展的一个重要因素,因发展而带来的强大客流往往超出了车站自身负荷,车站建设应根据城市的总体规划进行提前判断,并为车站提供具有发展性的客流分析。车站在客流容量及空间上均保持一定的包容性时,才能在如此快速变化的城市环境中,保持轨道交通的可持续发展及其城市价值。随着地铁线网的发展与完善,地下车站空间的多样性是地铁发展成熟的标志,地下车站从"城市机器"逐渐转型发展为城市综合体,车站设计的外部输入条件更为多样与复杂,多线换乘、多交通系统换乘、TOD交通综合体等都日渐成为地铁车站深入研究的课题,城市对未来的地铁车站的规划与设计提出了更多的要求与挑战。

The Way of Continuity (Underground Station)

1. The Continuous Space

The spatial form of urban metro was initially constructed in the form of single stations, which connect into a line through the tunnel, and the lines constitute the rail transit network via the interchange stations. The spatial form presents the trend of continuous development of "point", "line", "plane", "block" and "network".

The Form of "Point"

When the ancillary facilities on the ground such as ventilation pavilion, exit & entrance, cooling tower and small station hall on the ground combine with part of the service facilities, a hub of the station functions come into being, it is also the terminal and node of functional streamline and passenger streamline, demonstrating the features of "small", "scattered" and "multiple" in the urban spatial pattern.

The Form of "Line"

There are highly mobile and linear features of the vehicles or pedestrians passing spaces such as station tunnels, excavated platforms, pedestrian passageways and overpasses. The existence of these "lines" "links" the seemingly scattered hidden (underground) or visible (ground or elevated) or half-hidden and half-shown "points" together.

The Form of "Plane"

The scale of station hall is large for the multiline interchange station, though the absolute value of height is higher, the ratio of height and area is much smaller than ordinary stations. It also presents flattening spatial form and becomes planar space. Large-scale depot (base) occupies vase area with its urban spatial form in highly flattening feature. In the later stage, buildings atop will be developed to turn the entire depot (base) into the pedestal.

The Form of "Block"

The elevated station (whether on the road or on the side of the road), station platform and the station roof will give a more obvious sense of volume and become the visual focus of city view. Nevertheless, the sense of volume of some integral settings (equipment room on the road, station hall and platform are integrated) are more obvious.
Some underground stations are set in the hugeatrium-style underground space for external condition reasons, composing the gigantic underground "block".

The Form of Network

The Planning and construction of the entire lines in the area enables the extension and expansion of metro stations from line to network gradually covering the entire area. It is also different from most public transportation buildings (such as airport, dock and bus station etc.), only rail transit stations can realize this kind of continuous growth and eventually constitute the network.
This form of network can be perceived as completed or uncompleted. When the line is built all at once, it could be deemed as completed; while it is not the case and the line (Separation of Guangzhou Metro Line 2 & Line 8) will be separated to achieve the final form, this is a dynamic change and self-adjusting process.

If the connected part (ground and underground, tunnel and station) of all metro network is regarded as an integral whole, the volume and scale of this building is undoubtedly the largest and most extensive in the city. It involves almost the majority of areas in the city, and keep changing and expanding to become a "super building", which cannot be imagined merely from the scattered exit & entrance and ancillary facilities on the ground.
On the other hand, the appearance of this "super building" is "insignificant" from the partial view, that only the ventilation pavilion of a few square meters is seen.

Station Space Generated from Objective Conditions

The biggest difference between underground station and ordinary civil building is that the design and construction of underground station cannot be done on a separate and integral land. The location of the station is usually in the underground area of the arterial road with busy traffic and the underground pipe network is densely distributed.
The design also requires adequate consideration before hand to minimize the huge impact of demolition and construction on urban traffic and the surrounding environment, which is very challenging for the comprehensive technique and macro judgment of the designer.
In the meantime, the analysis and evaluation of the efficiency and commercial value of the surrounding passenger flow needs to be fully considered for station design to attain the optimal result through the comparison of multiple schemes.
In regard to station design, the construction conditions are taken into full account to turn the unfavorable factors into favorable ones so that the station space can be uniquely expressed. Take the Zoo Station of Guangzhou Metro Line 5 as an example, because the station is deeply dug to provide good conditions and form huge space for the partial public area of the station hall, creating space of distinctive characteristic for Zoo Station in combination with two "Y"-shaped structural columns.

2. The Continuous Time

Compared with other types of public buildings, metro station has clear dynamic features of mobility and procedure (other public buildings are more residing and static functional space such as hospital clinics and wards, school classrooms, high-rise office buildings, hotel restaurants and guest rooms, cultural center theaters etc. The public has certain purposes in these places and will stay for relatively longer period. The places feature terminalization). In contrast to other public buildings, metro station also has stronger purpose. Passengers often choose to go to a station or interchange at a station and then leave quickly, and usually stay for a few minutes showing the feature of high mobility. Therefore the place feature is proce during and continuous.

3. The Continuous Experience

Compared with the buildings on the ground, most of the space users in the station experience the "inside" space. Due to the space's one-way characteristic, passengers cannot sense the scenery and time changes of outside space. Contrasting with the "three-dimensional" experience of other public buildings, the pace of the metro station is mostly flat and interface-like, with strong planar nd scenario features, which can often deviate from the area where the station is cated, and formself-contained expression. Due to the similar characteristics of nost station public spaces, passengers cannot feel the changes on the ground or external environment while travelling through the underground stations. From ne angle of rail transit network of the entire city, the underground stations and ie interval sections form a continuous and integral space in which passengers onstantly switch in between. It is easy to lead to the continuous modeled xperience of "exit & entrance" — "passageway" — "station hall" — "station atform" — "car" — "tunnel" — "station" — "station hall" — "passageway" — "exit & ntrance". However, due to the homogenization of the scale and space form of fferent stations, the passengers share similar experience of lacking perception of urroundings while commuting on the underground rail transit, which makes the assengers feel bored. From the perspective of human experience, the station bace creates certain differences according to its own conditions. On one hand, he spatial form of the station combines various constraints such as environment, eology and structure to create differences of the station, on the other hand, the ation decoration incorporates regional disparity and expresses particular cultural tentions. With the goal of providing rich passenger experience in the metro :ation, the differentiated expression in terms of space, form and scale etc. of the :ation are sought to avoid the mediocre metro environment resulting from the variable standard.

The Continuous Growth

2005, Guangzhou Metro Construction Planning planned 15 lines of 619km; in 020, Guangzhou Rail Transit Construction Planning planned 19 lines of 817km; ne target of the metro construction gradually transits from auxiliary infrastructure in ervice for the city to leading development of the city; from single-line construction o multi-line implementation and then to the construction of the entire network; om construction of a single metro station to the upper and lower linkage of tegrated transportation hub and properties atop, and the scale of construction and e areas covered are also constantly being expanded and adjusted.
he dense underground urban metro stations provide active transportation onditions and huge pedestrian flow for these gigantic underground developments. elying on under ground stations, dynamic underground commercial streets sually evolve linking with business on the ground. At the same time, the compact rban development model also put forward new requirements and restrictions for he station, and the requirements for station adaptability are getting stronger and tronger.
he early construction mode of metro was large-scale demolition and construction parallel with urban construction, expanding outward in line with the space emand of metro construction, therefore the metro station form at early stage /as more of two-story standard. With the rapid construction and expansion f the city, the planning and implementation speed of the ground construction far higher than that of the underground space construction, resulting in the onstraint of underground station construction by the space form on the ground. 1 the meantime, due to the growing restriction for on-site construction caused by the change of urban traffic intensity, the station form is becoming more and more individualized in the trend of non-standardization complemented with standardization.

In the deep underground of Guangzhou, the metro lines cover most important areas of the city like entangling roots. Every metro station burdens the largest transportation capacity and passenger flow of the city every day, it has become the highest engaged building by the citizens. Other than common ground transportation buildings, underground metro station has no facade and shape and cannot give symbolic or direct visual impact to the city, but it carries unparalleled function and significance for the city. If the museums, libraries and art galleries belong to the upper Infrastructure of urban spirit and ideological, then the metro station is the underground base that supports upper Infrastructure of the entire city. Each station is like a low-key and high-efficiency precision machine component in the underground of the city, which is connected together through the integrated lines, as if a giant engine buried deep down of the city. The operation of the engine also drives the city running, and directly affects the citizens' activities, employment, living, business development, financial economy and urban expansion and extension.

As the vein of the city, the station is macroscopically connecting to different areas and even different cities; and microcosmically linking with the surrounding environment closely. It is a key factor for evaluating the efficiency and value of a station that if it connects and interacts with the surroundings. Therefore, it should be the important consideration in station design that how to keep progressive connection with the surroundings and adapt to the changes of the developing environment. Apart from the physical space connection such as the station passageways, attention should also be paid to the possibility of station changes to enable conditions for changes and fitting in surrounding continuous development in the future. In addition, the pre-judgment of passenger flow is also an important factor for the station to accommodate to the surroundings, since the huge passenger flow along with development often exceeds the capacity of the station, the station construction should be judged in advance according to the master plan of the city, and provided evelopmental passenger flow analysis. Inclusiveness should be maintained in space and passenger flow to keep sustainable development and city value of rail transit in the fast-changing urban environment. With the development and improvement of the metro network, the diversity of underground station space is a sign of maturity of metro development. The underground station slowly transforms from the "urban machine" to the urban complex. The external input conditions of the station design are more diverse and complicated, multi-line interchange, multi-transportation system interchange, TOD traffic complex etc. have gradually become the subject of in-depth study of metro stations. The city has brought up more requirements and challenges for the planning and design of future metro stations.

兼容之道
(高架及地面车站)

1. 气候与环境的兼容——气候环境的适应性

城市轨道交通的敷设方式一般分为地下敷设和地上敷设，根据城市总体规划，结合城市现状及地质、经济等多方面进行综合选择。城市轨道交通的建设成本高昂，地上敷设线路的工程成本远低于地下敷设。出于成本考虑，地上线路无论在欧洲还是亚洲的日本等国家的城市轨道交通系统中，都占据了不小的比例。

城市轨道交通由地下走出地面，尽管具有造价低、能耗小、工期快等优势，但同时也带来了一系列如噪声、景观等城市问题，值得深思。近年来，在各地经历了一批高架线路的建设后，北京、上海、广州等城市由于用地紧张，高架线路在新线规划中已基本停止了。但基于成本优势，地上敷设的高架线路在全国其他城市的轨道交通的快速发展中仍有一定比重，所以伴随高架线路的地面车站应当得到重视，为其更好地融入城市环境中做出努力。

为了解决上述高架线路的城市问题与成本优势的矛盾，地面车站需要兼顾城市环境中的方方面面，包括景观、环境噪声、站点人流换乘、城市发展等。一个车站往往兼容了城市的各种功能要求，也具有比普通公共建筑更为丰富的内容。长期以来，建筑学专业在轨道交通设计中处于配套地位。在地面车站或者高架车站的设计中，建筑学专业主导作用应得到体现，为车站寻求与城市的兼容之道。

兼容气候与环境

无论地面车站还是高架车站，都属于现代交通建筑，而建筑就必须是地区的产物，并且是具体环境与社会、人文、气候及地貌等不同要素相互作用的结果，与城市环境融合。

从地域的气候环境来说，高架车站站台基于乘客停留时间较短，采用机械排烟和自然通风，一般只在站厅设置空调系统。为提供舒适的站台乘车环境，在车站的建筑设计中，必须根据车站所处区域的气候区域特征，考虑车站自然条件的利用。采光、通风、遮阳及挡雨几方面的功能，是车站建筑设计中需要充分考虑的因素。同时也应该结合所处具体自然环境或者城市环境，综合考虑体量、形式、材料以及色彩。比如在南方地区，由于气候多雨、炎热，阳光充足，人们崇尚室外活动，亲近自然。在具体高架车站的设计中，更多地强调通风和挡雨，同时车站空间与室外相互渗透，站台空间与室外的边界模糊，融为一体，自然而然地形成车站通透、轻巧、明快的风格特征。

从车站所处的环境而言，车站的形式应该挖掘地域的文脉、传统以及地区发展规划，让车站在兼顾现代交通建筑的形式语言外，符合环境特征，成为从环境中"生长"出来的建筑。因此，在建筑造型设计上，应以简洁的形象为主调，通过周边环境、地域特色、城市文脉及城市规划等方面深度挖掘，提取出属于车站应有的建筑元素及造型，使车站建筑在满足自身使用功能的前提下，形成自我的形象特征、造型标识及一定的环境元素。

2. 城市空间的景观兼容

高架线路的车站往往以地面车站或者路中高架车站的形式出现，无论哪种形式，其最终结果都将对城市景观产生积极或消极的影响。能否处理好城市道路，街关系、尺度形式的关系，很大程度上决定了车站的成败。

作为现代交通建筑首先需要表达出功能的性格特征，传达出建筑的速度感和效感。车站在城市中应该展现出明快、开放、流畅的时代气息，并最终以强烈的交通建筑标识性呈现给大众。在人们出行的过程中，能够通过其鲜明的形象特征，便捷地找到车站。另外，在车站的建筑设计上，还应注重塑造车站的场所神，形成车站的抵达感。所谓抵达感，即利用车站建筑造型元素、外部空间设计等手段对车站周边环境形成一定的标志性，如广场空间的营造与建筑特征的塑都会对环境产生相应的影响，使人们在进站乘车前形成一定的仪式感，同时也是建筑自身对其周边环境所作出的回应。

车站建筑的造型设计须通过自身与环境的互动，把握车站功能的表达及环境素，才能在设计上形成车站的识别性，才能强化车站的功能特征、环境特征及所氛围。

另外，轨道交通车站建筑日常的使用频率非常高，同时是与人直接接触的交通间载体。因此，车站建筑在设计上应充分考虑人的需求与空间感受，除了在功上，如无障碍设计、导向、节能、广告等方面提升车站的服务质量外，在空间造型、细节上同样需要注重人的切身感受，营造出良好的印象，体现出精细的设计品质。在具体的建筑设计上，应该在建筑的细节中，精心策划，从文化历史、材料特征等方面进行细致的挖掘，塑造良好的建筑细节特征，结合建筑式、功能，提升车站的整体设计水平，创造出高品质的车站空间载体。如在细设计中，从车站的地域文化中提取一定的文化元素，运用到具体的细节设计上体现出区域文化背景。这在一定程度上丰富了车站的设计内涵，让车站作为城景观的重要组成，有更高的品质体现。

兼容可持续发展

市轨道交通作为解决城市可持续发展的重要环节，在具体车站或沿线的建筑物
计上，更应将可持续发展的理念落到实处，从节能减排、低碳环保方面考虑，
升建筑设计的内涵及车站建筑的效率。在条件允许的情况下，建议车站顶棚采
光伏玻璃，白天充分利用自然光并储存太阳电能，且在夏季光照强烈的条件下
收太阳光，起到一定的遮阳效果，夜晚用于车站照明并使用新型节能型照明设
和照明方式。另在场地空旷的区域，在条件具备的前提下，也可采用风能、太
能发电设备，以达到资源再利用的效果。根据不同功能区采用不同照度标准。
效控制广告用电量。但往往主动式节能措施在运用上较为繁琐，对具体的建筑
型影响较大，因此在设计上，应充分考虑节能措施与车站建筑造型结合的可
，使设备与造型有机结合，形成统一的形象，避免车站中过多出现繁多的设
，破化立面。

兼容客流与城市发展

架线路一般选择在城市用地宽松的近郊区域，站点间距较大，服务的范围相对
区站点更广，因此，高架车站在总体规划时，更应注重客流吸引与换乘需求。
架车站的乘客，由于服务半径往往超过1000m，超出了正常的步行距离，更多
要借助其他交通工具抵达车站，如自行车、公交车、出租车、私家车及摩托车
，车站地面需要提供更多的广场、停车及换乘空间，一方面满足客流到达的便
性，另一方面实现轨道交通与其他交通工具的衔接。
此同时，近年所倡导的TOD城市发展模式强调轨道交通对城市发展的引导，高
车站的站位选择往往对日后片区的发展起着积极的作用，引导周边开发，而土
开发也为轨道交通带来更多的客流，相互促进。所以，今后高架线路的车站，
其是独立地块的车站，应该更多地考虑复合型开发，将轨道、物业、换乘等
能进行综合开发，通过中庭等空间进行高效的联系，并可考虑采用分期开发模
，将车站提前完成。随着城市不断发展，客流不断提升，其他功能在车站的基
上不断"生长"，不断完善。在这样的模式下，既可发挥轨道交通高架线路的
本优势，也可在土地开发上取得高效的利用。

The Way of Compatibility (Elevated and Ground Station)

1. The Compatibility of Climate and Environment — Adaptability of Climate and Environment

The ways of laying urban rail transit are generally divided into underground laying and ground laying, which is the comprehensive choice according to the urban master plan in combination with the city's current situation, geology and economy etc. Due to the high construction cost of urban rail transit, the engineering cost of laying on the ground is much lower than that of underground, in consideration of that, metro line on the ground has taken certain percentage in the urban rail transit system in Europe and Japan and so on.

Urban rail transit runs from underground to the ground, despite the advantages of low cost, low energy consumption and shorter construction period, it also brings a series of urban problems like noise and landscape which is worth pondering. After the construction of elevated lines in various cities in recent years, the planning of new elevated lines is basically suspended in Beijing, Shanghai and Guangzhou etc. for land shortage reasons. However, due to cost advantage, elevated line laying on the ground still takes up certain ratio in the fast development of rail transit in other cities across the country, hence attention should be paid to the ground stations of elevated line for them to better integrate into urban environment.

In order to solve the contradiction between the urban problem and cost advantage of the above-mentioned elevated line, the ground station needs to take into account all aspects of the urban environment including landscape, environmental noise, station passenger interchange and urban development etc. A station comprises various functional requirements as well as richer content than ordinary public buildings. For a long time, architectural design has been of the supporting function in rail transit design, while it should play a leading role in ground station or elevated station design to seek for compatibility with the city.

The Compatibility of Climate and Environment

Both ground station and elevated station are modern transportation buildings, which bound to be the regional products, and as well the interaction result of different elements including environment, society, humanity, climate and geomorphology which integrate with urban environment.

As for the geographical climate environment, based on passengers' short staying, the elevated station platform applies mechanical smoke exhaust and natural ventilation, and air conditioning system is only equipped in the hall. To provide comfortable platform environment at the station, utilization of natural condition should be taken into account according to the climate characteristics in the station architectural design. Functions including lighting, ventilation, sunshade and rain protection are factors to be fully considered in the design of station building. Simultaneously volume, form, material and color should also be comprehensively considered in combination with the specific natural or urban environment. For example, it is rainy, hot and sunny in the southern region, where people like outdoor activities and getting close to the nature. In the design of the elevated station, ventilation and rain prevention are focused. Also the station and outdoor space penetrate and integrate into each other by obscuring the boundary, developing the transparent, light and bright style of the station naturally.

In terms of the environment in which the station is located, the form design of the station should excavate the culture, tradition and development planning of the area. The station should be the building "growing" in the environment conforming to the environmental characteristics including the form language of modern transportation building. As a result, the building form design should be themed on the concise image, deeply exploit the surroundings, regional characteristics, city culture and urban planning etc. to extract its exclusive building element and shape, forming its own image feature, image icon and environmental elements on condition that the station building can satisfy the usable functions.

2. The Landscape Compatibility of Urban Space

The station of elevated line often appears in the form of ground station or elevated station on the road, whichever form it is, the final result will have positive or negative impact on urban landscape. The success of the station depends on the proper handling the relationship of urban road, street relationship and scale form to a large extent.

As the modern transportation building, the functional characteristics need to be expressed in the first place to convey the building's sense of speed and efficiency. The station should display the fast, open and smooth contemporary spirit and ultimately present a strong iconic image of transportation building to the public. During people's travel, the station can be found conveniently with its outstanding image features. Moreover, the architectural design of station should also mould the place spirit and create the station's sense of arrival. The meaning of sense of arrival is to utilize the architectural modeling element, exterior space design etc. to create icon for station surroundings, for instance, the buildup of square space and building feature will exert corresponding influence on the environment, and give a sense of ceremony before passengers entering the station, it is also the response to surrounding environment from the building itself.

Through the interaction between the environment and the shape design of the station building, the master of expression and environmental elements of the station function can reveal the identification of the station, and then reinforce the functional feature, environmental feature and station atmosphere.

In addition, being the transportation space carrier which is in direct contact with people, the building of rail transit station has a very high usage rate daily. Therefore people's needs and space perception should be fully considered for the design of station building. Besides in functional aspect, such as barrier-free design, guiding, energy saving and advertising etc. will enhance service quality of the station; people's personal experience should be taken care of in spatial, formative and detailed aspects to create pleasant impression and reflect refined design quality. As for the specific architectural design, it should be carefully planned in architectural details, and excavated meticulously in culture, history and material feature etc. to make nice architectural details, combine architectural form and function, enhance overall design level of the station, and create high-quality space carrier for the station. In detail design, cultural elements are to be extracted from regional culture of the station and applied to specific detail design to reflect the cultural background of the area. This enriches the design connotation of the station to a certain extent, and makes the station an important part of city landscape and shows higher quality.

The Compatibility of Sustainable Development

Urban rail transit is an important part of solving the sustainable development of the city. The concept of sustainable development should be implemented in the design of specific stations or buildings along the line. Considering energy saving, low carbon and environmental protection, the significance of building design and efficiency of station building should be upgraded. When conditions permit, it is suggested to use photovoltaic glass in the station roof to make full use of natural light and store solar energy during the daytime, absorb sunlight in the strong summer sunshine to achieve a certain shading effect, and apply new energy-saving lighting equipment and lighting methods at night. With proper conditions, wind power and solar power generation equipment can also be utilized in open areas to achieve the result of resource recycle. Electricity consumption can be effectively controlled according to different illumination standards in different functional areas. However active energy-saving measures are often troublesome, which will affect greatly on specific architectural shapes. Therefore the possibility of combining energy-saving measures with the building shape of the station should be fully considered in design process, so that the equipment and shape can be organically integrated to form a unified image, and avoid excessive equipment on the station to spoil the facade.

The Compatibility of Passenger Flow and City Development

The locations of elevated lines are generally selected in the suburban areas where lands are more vacant, distance between stations is longer, and scope of service is wider than urban areas. Therefore, more attentions should be paid to passenger flow attraction and interchange requirement in the master planning of elevated station. As the service radius is usually more than 1000m, which exceeds normal walking distance, passengers of the elevated station depend more on other means of transportation to reach the station, such as bicycle, bus, taxi, car and motorcycles etc. The ground over the station needs to provide more squares, parking and interchange spaces, to satisfy the convenience of passenger flow, and realize connection of rail transit and other means of transportation.

In the meantime, the TOD urban development model being advocated in recent years emphasizes the guidance of rail transit for urban development. In most cases, the selection of elevated stations location will have active impact on the development of the area and guide the peripheral development. As the land is developed, it also brings more passenger flow to the rail transit and plays a mutually promotion role. Hence in the future, the stations of elevated lines, especially those with independent land plots, should consider compound development more, develop comprehensive functions such as metro, property, and interchange, make efficient links through space like the atrium, and attempt to complete the station construction ahead of schedule via development by stages. As the city develops, the passenger continues to increase, and other functions keep "growing" and improving on the basis of the station. In this mode, it not only takes advantage of cost-effectiveness of metro elevated lines, but also high efficiency use of land development.

文化之道
（车站公共区装修）

1. 个性与共性

共性与个性是车站装修设计永恒的主题。

车站建筑作为城市公共建筑的主要设施空间，为使用者提供清晰的识别性是其最为主要的功能特征。高度明确的导向性在封闭的地下空间中是为公众提供安全高效使用体验的前提。除了车站导向牌等硬件标配之外，公众对车站空间场所的自我识别与文化认同也日趋强烈。从广州市轨道交通一号线至今，车站的识别性的理解与实践，共性与个性，纠结与反复，相互催生出多样化的设计成果。

车站设计模式从一站一景（广州市轨道交通一号线），一站一色（广州市轨道交通二号线，二、八号拆解工程线，三号线），一线一材（广州市轨道交通四号线的地下玻璃和高架钢材铝板），到一线多材（广州市轨道交通五号线、六号线的顶棚不锈钢），一线一景（线网概念，广州市轨道交通四号线南延段，九号线，七号线，十三号线）进行了各种模式的尝试。

设计特性表达也经历了从个性表达（广州市轨道交通一号线）、共性表达（广州市轨道交通二号线，二、八号拆解工程线，三号线）、个性与共性混合（各线独立表达），到个性与共性融合（共同表达），以及最终在线网框架下，各线主题个性化，车站表达模式以标准站、重点站及文化站的组合模式进行设计的过程。

从城市线网与为公众服务的角度出发，轨道交通建筑有必要建立一种区别于城市环境的个性；而从每个乘客对地铁系统的识别及交通导向的角度出发，又需要建立一种具有高度特征化的个性；同时作为掌管轨道交通建设及运营的组织，也有必要建立具有明显VI系统的企业文化体系。

个性往往会在轨道建设的早期、在站点数量较少时有强烈的表达意愿；随着建设强度的提升，适应建设量的巨大考验，"工业化，模块化，标准化"的共性必然成为主题；而当站点建设数量积累到一定程度，颜色、材料等较易取得明显设计效果及个性识别的手法走向末路的时候，对个性的彰显又重新得到重视，不过个性的表达暂时还没有形成主流，只是在某些"文化重点站"进行设计的尝试。

2. 技术与艺术

受制于车站交通主体功能及乘客安全需求，车站室内装修设计长期停留在技术设计为核心的层面。装修界面的处理仅满足于对乘客使用界面和管线空间界面进行整合。设计层面更多地关注施工快捷模式及多方施工的需求，造价控制与招标式的限制，运营后期维护保养的实际需求等，上述需求基于对现实的顾虑，催了"工业化，模块化，标准化"的技术性指导原则。在地铁高强度建设时期，述模式也确实有效解决了建设期间的相关问题。

近年来，从宏观层面的发展政策释放出对文化及精神层面的重视和强化的信息大众对美好生活向往，可以成为设计追求的目标与方向。作为公众高密度的所，随着对界面材料的使用到达极限，不可避免地会造成大量地铁公众场所呈"多城一面"的单一面目，形成大众在视觉上的审美疲劳。地下和室内的封闭境与外部完全隔离的空间，客观上带来方向感迷失和乘客心理的困惑，加上大单一界面的室内场所数量的增加，已带来大众对场所识别性的降低和弱化，对站交通主体功能造成影响。长期在单调的视觉体验中穿行，方向感的迷失容易成乘客心理状态的波动，也不可避免地对乘客安全造成影响。

基于上述现状，客观上需要在地铁车站形成重点与差异性的空间界面，部分处城市重要地段或具有历史文化背景的区域，也要求相应地段的公共建筑对这些境元素有所回应或表达。车站装修设计的艺术化表达已成为越来越迫切的需求。伴随社会意识形态重要性的逐步增强，车站作为民众最常用的公共空间，部分有历史背景的站点有了文化表达的诉求，车站装修便进入第三阶段——文化形介入模式，包括线网文化设定，文化墙、文化站、特殊空间站等不同形态的入。在线网文化设定中，不同线路根据其地域特征设定相应的文化主题并以线为单位进行主题化设计，在城市线网中线路数量越来越多的情况下以线路为单制订了识别性规划策略。从文化墙进入地下车站开始，地铁交通空间开始进入单向空间发展为多向空间的新时代。车站除了承载日常的交通功能外，还可以担企业、城市、线、站点等多元的文化。多元文化表达为车站空间创造了更多可能，部分重要特殊文化背景的车站可从文化墙二维界面发展为空间与场景式三维和四维的空间表达。

文化站文化介入分为三个层次。第一为看山是山，以画写实。看见文化主题态，就单纯地通过剪贴画和黑板报的形式，生硬地拼贴于墙面上，如商业电影生硬的广告植入手法，为其车站强加出市民能看得出的"文化"。第二为看山是山，以实写实。看见文化主题有什么元素，就把元素复制入车站装修元素中最终建筑却变味了，变成了四不像，变得有些庸俗，看山不是山，看车站也不"站"。第三为看山还是山，以虚写实。不刻意照搬文化主题的元素体现文，但能将文化的内涵与意境和交通建筑属性恰如其分地融合在一起，既是文化是交通，既是"山"，也是"站"。精神层面与场所感的文化体验像空气一样入车站的每一个角落，成为一个真正意义的"文化站"。如广州市轨道交通五线动物园站清水混凝土Y柱高大空间表达自然界的原始与粗犷，广州市轨道交十三号线南海神庙站"海不扬波"的祭祀场景体验，广州市轨道交通四号线南段的南沙客运港站作为丝路起点的"宝船起航"壮阔场景，都体现出文化设计文化展示墙到文化展示区再到整体文化站的思想提升过程。

材料与约束

于目前材料使用及消防安全规范的限制，目前在轨道交通站点室内的材料基本是三大主材。吊顶以金属（铝合金）为主；墙面受香港地铁早期的影响，以搪钢板为主，后期从广州市轨道交通四号线发展出夹胶钢化玻璃墙板的使用；地从开始就以石材为主；室内栏杆及室内设施（商铺、票亭等）以不锈钢作为主选材。

比一般民用建筑室内装修多元化的材料选择，地铁对装修材料的选择主要考虑下因素：一是能适应地铁高强度及高速的建设需求，最好在本地或国内具有较强大的产能；二是适应地铁工期较短的建设模式（可考虑室内多个界面同时施）及招标模式（以材料为主的招标模式可取得较高的性价比），国内有多家供商可供选择；三是在未来的运营及维护中，具有良好的保洁、耐磨、抗破坏及维护的特性。综合考虑上述需求，目前可供选择的建筑材料确实不多。

料选择的限制，客观上也对车站的室内装修设计形成约束。

顶要考虑方便运营期间对设备管线的检修，从消防角度要考虑足够多的透空性足排烟的要求；墙板要考虑广告灯箱的模数，墙面设备的安装开启；地面则要虑各种地面设备开口，楼扶梯洞口收边，现场良好的二次加工的特性；对吊顶形态设计还要考虑各种系统设备安装，导向安装等是否能与吊顶的形式及安装匹配。

The Way of Culture
(Decoration of Station Public Area)

1. Individuality and Universality

Individuality and universality are the eternal theme of station decoration design. Station building as the main facility space of urban public buildings, its most important functional feature is to provide users with clear identification. Highly specific guiding is the precondition for providing safe and efficient experience for the public in closed underground space. Besides the standard equipped hardware such as guide boards, the public's self-identification and cultural identity of the station space is also growing stronger. From the construction of Guangzhou Metro Line 1 to the present, diverse design results have been generated by the understanding and practice, individuality and universality, entanglement and repetition of the identification of the station. Various attempts have been applied on station design modes from one station one scene (Guangzhou Metro Line 1), one station one color (Guangzhou Metro Line 2, Separation of Line 2 & Line 8, Line 3), one line one material (Guangzhou Metro Line 4 underground glass and elevated steel aluminum plate), to one line multiple materials (Guangzhou Metro Line 5 and Line 6 stainless steel ceiling), one line one scene (network concept, Guangzhou Metro Line 4 South Extension, Line 9, Lines 7, Lines 13).

The expression of design characteristics has also gone through individuality expression (Guangzhou Metro Line 1), universality expression (Guangzhou Metro Line 2, Separation of Line 2 & Line 8, Line 3), individuality and universality mix (independent expression of each line), to individuality and universality integration (co-expression), eventually each line has individualized theme, and the station expression mode is designed in a combined way of standard station, key station and cultural station under the network framework.

From the perspective of urban network and public service, it is necessary for rail transit buildings to establish the individuality that is different from the urban environment; from the perspective of each passenger's identification of metro system and transportation guiding, it is necessary to establish highly characterized individuality; meanwhile for an organization that controls the construction and operation of rail transit, it is also necessary to set up corporate culture system with an obvious VI system.

Individuality tends to have strong willingness to express when the number of early sites is small in the track construction; with the increase of construction intensity, in order to adapt to the massive construction workload, the universality of "industrialization, modularization and standardization" will inevitably become the theme; when the number of site constructions is accumulated to a certain extent, and approaches that can easily achieve obtain obvious design effects and individual identification such as color and material are coming to an end, individuality is again highlighted, however the expression of individuality has not yet formed the main stream, instead, it is only by experiments of designing some "cultural key stations".

2. Technology and Art

Subject to the requirements of transportation function and passenger safety of the station, the interior decoration design has long stayed around the core of technical design. The processing of the decoration interface can only satisfy the integration of passenger's user interface and pipeline space interface. On design level, more attentions are paid to requirements of construction shortcut mode and construction by different parties, restrictions of cost control, and actual needs for maintenance in later operation etc. which are based on consideration about reality, resulting in the technical guideline of "industrialization, modularization and standardization". During the high-intensity construction period of metro, the above model did effectively solve the related issues in the course of construction.

In recent years, the message of emphasizing and strengthening the cultural and spiritual aspects has been released from macro-level development policies. People longing for better life can be the goal of design pursuit. As the public place of high-density, along with the limit of interface materials usage, it will inevitably result in the monotonous appearance of "identical images" in many metro stations and the visual aesthetic fatigue of the public. The underground and indoors closed environment and space completely isolated from the outside objectively causes disorientation and confusion of the passengers. Moreover, the increase in the number of indoor space of single interfaces has brought about reduction and weakening of the public's identification of the place, and impacted on the main function of station transportation. Disorientation is likely to cause fluctuations of the passenger's psychological state and impact on passenger safety unavoidably after long-time traveling in monotonous visual experience.

Based on the above-mentioned status quo, it is objectively necessary to generate the key and different spatial interface in the metro station, and those located in important urban areas or areas with historical and cultural backgrounds are also required to express or respond to these environmental elements with public building in the corresponding areas. The artistic expression of station decoration design has become an increasingly urgent need.

With the increasing importance of social ideology, the station is the most commonly used public space. Some stations with historical background have the appeal of cultural expression, and the station decoration enters the third stage— the cultural pattern intervention mode including different pattern intervention of network culture setting, cultural wall, cultural station and special space station and so on. In the network culture setting, different lines have themed design according to the cultural theme corresponding to their regional characteristics based on the lines, with the main purpose of creating the identification planning strategy based on the lines responding to the growing number of urban network lines. Starting from the cultural wall into the underground station, the metro transportation space began to enter a new era of developing from one-way space to multi-way space. In addition to daily transportation functions, the station can also undertake the diverse culture of enterprises, cities, lines and stations. Multi-cultural expression has created more possibilities for the station space, some important and special stations with cultural background can be developed from the two-dimensional interface of the cultural wall to three-dimensional and four-dimensional spatial expression of space and scenario.

The cultural intervention of cultural station is divided into three levels: firstly is "the mountain is the mountain", to design what is seen, just directly piece cultural theme pattern together on the wall simply by means of clip art and blackboard drawing, like the stiff product placement in commercial films, imposing the visible "culture" on the station; secondly is "the mountain is not the mountain", to design what it is, copy atever element is found in the cultural theme in the station decoration element, d eventually lead to failure copycat, making the building neither fish nor fowl and e station unaesthetic; thirdly is "the mountain is only the mountain", to design what contains. The elements of the cultural theme are not deliberately imitated to reflect e culture, instead, the cultural connotation and artistic conception can properly egrate in the transportation building attributes. It is both culture and transportation eans, both "mountain" and "station". The cultural experiences of spirit and the nse of place are implanted in every corner of the station like the air. The station is coming a "cultural station" in the real sense. For instance, the huge space of as-st-finish concrete Y-pillar at Guangzhou Metro Line 5 Zoo Station showing the mitiveness and ruggedness of the nature; the worship scenario experience of "the m and tranquil sea" at Guangzhou Metro Line 13 Nanhai Temple Station; the grand enario of "the treasure ship setting sail" at the origin of the Silk Road —Guangzhou etro Line 4 South Extension Nansha Passenger Port Station. All these have dicated the thought upgrade process of cultural design from cultural display wall to tural display area, and then to overall cultural station.

Materials and Constraints

view of the current restrictions on material utilization and fire safety regulations, at esent there are basically three main materials for interior design of metro station. etal (aluminum alloy) is used for the ceilings; enameled pressed steel is applied the walls influenced by the early construction of Hong Kong MTR, however minated tempered glass wall panel was developed since Guangzhou Metro Line 4 later stage; stainless steel is utilized for indoor railings and facilities(shops and ticket oths etc.).

ompared with the diverse material choices for interior decoration of general civil ildings, the following factors are principally considered for the choice of metro coration materials: firstly, it should adapt to the high-strength and high-speed nstruction demands, so preferably it could have larger production capacity in the y or in China; secondly, it should fit in the construction mode of short construction riod (construction simultaneously on multiple interior interfaces can be considered), ich many selectable domestic suppliers are available; thirdly, it should feature easy eaning, abrasive resistance, damage resistance and easy maintenance. In overall nsideration of the above requirements, there is really not much to choose from the rrent building materials.

e limitation of material selection also objectively constrains the interior decoration sign of the station. Selection of ceiling material should consider convenient aintenance of equipment pipelines during the operation period, and enough air rmeability to meet the requirements of smoke exhaust from fire protection angle; lection of wall panel material should consider modulus of advertising light box d installation and opening of wall equipment; selection of ground material should nsider various ground equipment openings, close sides of escalator openings, and od suitability of reprocessing on the site; shape design of ceilings also requires to nsider installation of various system equipment, and matching of guiding installation c. with the form and installation of the ceiling.

多元之道
（轨道交通配套建筑及景观）

1. 分散与聚集

通常所说的轨道交通地面建筑，是指地面出入口、无障碍电梯、风亭（包括区间风亭）、冷却塔、紧急疏散出口、区间变电所及其各种功能和形态的组合与搭配；同时还包括车辆段内的各单体建筑、检修厂房、停车列检库及运用库、综合楼、地铁公安综合楼、监控中心等。

从城市整体角度而言，上述各种地面建筑形态是分散的，并各自以不同的方式与对应的城市环境融合。或是位于独立的城市地块中，与周边的建筑保持足够的规划距离或环保间距；或是与地面交通设施（如公交站场、高铁、机场）结合设置，成为一种枢纽型的交通设施；或是与地面其他市政设施（公厕、变电所等）结合，形成小型的市政综合体。比较常见的还是结合地块的TOD开发，各种地面建筑融入TOD开发建筑群体。

多数车辆段由于征地及运营功能配置的要求，大多位于城市的边缘地区，在轨道交通的网络中也处于一种分散配置的状态。车辆段内部，相对于大体量而扁平化的检修厂房、停车列检库及运用库，其他建筑单体的形态也散布在主体厂房的周边。即使目前很多车辆段后期进行了上盖开发，虽然厂房的体块"消失"了，但余下的各建筑单体仍然独立并分散设置于地块内。

换个角度看，这些分散的建筑单体，其实都是一个整体的部分。如果把所有地铁网络连通的部分（地上及地下，隧道及车站）看成一个整体建筑的话，上述这些地面建筑都是其中的每个有机组成的元素；通过不断延伸的轨道网络，上述各分散的个体却以一种难以察觉的方式聚集在一个整体之内。

这种聚集从最小单元的组合（出入口与无障碍电梯，出入口与风亭，风亭与风亭，出入口加风亭及冷却塔等）到地面小型枢纽，综合体到结合地块的TOD开发，聚集的形态与强度呈现多元及逐级增强的变化。

2. 识别与融合

与车站装修的个性与共性相类似，地面建筑存在识别与融合的选择。当地面建筑独立设置于城市地块或每种相对孤立的城市环境中，特别是地面建筑的体量相于周边环境中的建筑或体块相对较小的时候，多数情况下，有必要强化地面建筑的个性，尤其是车站出入口、无障碍电梯等乘客直接使用并需要第一时间识别元素；当地面建筑和其他小型交通设施或市政配套设施结合时，其特征表达更是要表达一种混合的形态，体现对不同元素的包容性，强调自身内部的融合，内至外催生识别性；当地面建筑以TOD开发进行时，识别性可能要降至最低建筑形态更多的是与整体开发融合，但仍然需要保持对普通乘客的导向标志的别；而在车辆段上盖开发中，这种识别与融合需达到一种平衡和协调，检修房、停车列检库及运用库等扁平大体量与上盖物业形成融合，其他综合楼、地公安综合楼、监控中心等建筑单体由于功能及管理的要求，与其他上盖开发形独立的布局，也由于设计导向的差异形成相对的识别。

3. 形态与内涵

基于不同的技术条件、气候条件及城市环境、文化背景，各地轨道交通的附属面建筑呈现的外部形态有明显的差异性与多元化特征。总结目前各地的基本模大致有以下几类：一城一景，以青岛为代表；一城多景，为大多城市地铁所用；混合模式一城一景加一城多景，出入口模式相对统一固定，风亭及其他附设施形态相对多变。

各地轨道交通在长期的建设中均不约而同地将出入口作为地铁的标志性建筑，同时把出入口作为城市的"文化地标"，要求对所在城市的文化内涵有所呼应体现，追求表达寓意的层次丰富，以及与外部形态的简洁清晰，构成完整统一市视觉焦点，从而在城市繁杂的视觉背景中起到对公众有效而积极的引导和标□作用，同时也能呼应城市文化内涵最为普遍公认的价值取向。

□种程度上讲，地面建筑的形态不仅要联系和体现城市的文化价值，也要响应轨□交通企业的文化。从企业的VI系统到具体车站管理的模式，从地面建筑外部材□的选择，到相关地面建筑招标模式的确定，其实也是无时无刻不与企业人文精□相关联的。

车辆段综合楼造型设计上，将建筑的外观建立在其严谨的功能逻辑上，以理性□节制的理念表达轨道交通后勤办公建筑简约而现代化的气质，通过开窗方式与□局应对内部不同的使用需求，自然生成具有功能识别性的建筑外观，使建筑创□出一份属于其自身的独特的现代主义理性美学。

□地城市轨道交通的建设都经历了一个摸索的过程。形态与内涵，一内一外，从□期对外观造型重视单一的视觉效应，注重由外而内，到后期注重与标准化的技□结合，以及建设的效率性，偏重由内至外，再到现状，既要重视对公众的形象□导和传达，更要注重与公众深层次形成心理相应，强化乘客使用的持续性和识□的准确性。同时也注重在不同地段城市环境中采用多元化实施策略，强调内外□修；不管是"隐"还是"现"，都能很好地与城市环境和文化背景相得益彰。

The Way of Diversification
(Metro Auxiliary Building and Landscape)

1. Scattering and Concentration

The so-called rail transit ground building usually refers to the combination and matching of various functions and forms of ground exit & entrance, barrier-free lift, ventilation pavilion (including interval section ventilation pavilion), cooling tower, emergency evacuation exit and interval section substation; it also includes all individual buildings in the depot, maintenance workshop, parking inspection and operation workshop, complex building, metro public security complex building and control center etc.

From the perspective of the city as a whole, the above various ground building forms are scattered and integrated with the corresponding urban environment in different ways; they either locate in the independent urban land and maintain adequate regulated distance or environmental protection distance with the surrounding buildings; or integrate with ground transportation facilities (e.g. bus stop, high-speed railway and airport) into transportation hub; or combine with other municipal facilities on the ground (public toilets, substations etc.) into small municipal complex. TOD development with the land is more common, and a variety of ground buildings are incorporated into the TOD development building group.

Most of the depots are located in the edge areas of the city due to the requirements of land acquisition and operational function allocation in a state of scattered allocation in the rail transit network. Inside the depot, compared with the large-volume and flattening maintenance workshop, parking inspection and operation workshop, other single buildings are also scattered around the main building. Despite the buildings atop are developed for many depots at the later stage now, although the block of the plant "disappears", the remaining building are still independent and scattered in the land.

From another angle, these scattered buildings are actually a part of the whole. Regarding all connected metro networks (ground and underground, tunnels and stations) as a whole building, the above-mentioned ground buildings are the elements of organic components; these scattered individuals concentrate in one whole in an imperceptible way through the extending rail network.

This concentration ranges from the combination of the smallest units (exit & entrance and barrier-free elevators, exit & entrance and ventilation pavilion, ventilation pavilion and ventilation pavilion, exit & entrance plus ventilation pavilion and cooling tower etc.) to the small-scale hub on the ground, from complex to TOD development combining the land. The form and intensity of concentration shows the diverse and progressively increasing change.

2. Identification and Integration

Similar to the individuality and universality of station decoration, there is choice for identification and integration of building on the ground. When the ground building is set up independently in the urban land or the comparatively isolated urban environment, especially when the volume of the ground building is relatively smaller comparing to the building or the block in the surrounding environment, it is necessary to strengthen individuality of the ground in most cases, especially for the elements directly used by passengers requesting identification in the first place such as station exit & entrance and barrier-free elevation etc.; when the ground building and other small transportation facilities are combined with municipal facilities, its characteristic are more expressed in a mixed form, in the inclusiveness of different elements, in the emphasis on the internal integration and identification from the inside to the outside; when the ground building is developed by TOD, the identification may be minimized and the building form is more integrated with the overall development, but it still needs to maintain the identification of the guiding signs for ordinary passenger in the development of buildings atop the depot, this identification and integration achieves balance and harmony, buildings atop integrates with flattening and large volume of maintenance workshop, parking inspection and operation workshop etc. As per the functional and managerial requirements, other complex building, metro public security complex building and control center etc. adopts independent layout of the development atop, which also constitutes relative identification due to the differences in design guidance.

3. Form and Connotation

Based on different technical conditions, climatic conditions, urban environment and cultural background, the external forms presented by the affiliated ground buildings of various rail transits have distinctive features of differentiation and diversity. The current basic patterns of different localities are summarized into the following types of modes: One scene in one city represented by Qingdao; multiple scenes in one city adopted by most urban metros; mixed mode (one scene in one city plus multip scenes in one city, the entrance and exit mode is relatively unified and fixed, and the ventilation pavilion and other ancillary facilities are relatively more changeable).

In the long-term construction of rail transit, the entrance and exit is invariably taken as the symbolic building of metro in different cities, and the "cultural landmark" of the city as well, requiring to echo and reflect the cultural connotation of the city, pursue the rich levels of expression meaning and simple and clear external form to create

...e complete and unified urban visual focus. Therefore it can effectively and actively guide and indicate the public in the complicated visual background of the city, and at the same time respond to the most universally recognized value of urban cultural connotation.

To some extent, the form of the ground building not only needs to connect and manifests the cultural value of the city, but also respond to the culture of the rail transit company. From the company's VI system to specific station management model, from material choice for ground building exterior to confirmation of bidding mode of ground building, in fact it is also associated with the company's humanistic spirit at all times.

In the design of the shape of depot complex building, the building facade is built on the rigorous functional logic, to express the simple and modern quality of the logistics office building of rail transit with the concept of rationality and restrain, to cope with various internal usage requirements through the layout and window opening style and naturally generate the building facade of functional identification, giving the building unique rational aesthetics in modernism of its own.

The construction of rail transit in many cities has experienced a process of exploration. The form and the connotation are respectively inside and outside. From the emphasis on sole visual effect for building facade at early stage, outside-in was focused; to the standardized technical combination and efficiency of construction at later stage, inside-out was regarded as important; up to now, importance is attached to image guiding and conveyance and more to the in-depth psychological response of the public, enhancing passenger's continuous use and identification accuracy. Meanwhile attention is paid to adopt diversified implementation strategies in different urban environments and stress on giving equal importance to both inside and outside; whether it is "hidden" or "visible", urban environment and cultural background can complement well each other.

创新之道
(广州海珠环岛新型有轨电车试验段)

1. 差异与统一

有轨电车在交通定位上,是作为城市地铁交通系统的补充型交通系统,受交通运力和专用路权的需求限制,较适合设于小片区和新城区的内部交通衔接,并接入城市地铁大线网,使城市的轨道交通实现"小环接大环"环环相接的辐射型网络。作为一种小级别的轨道交通设施,有轨电车与常规城市轨道交通有很多相似之处,包括车辆、轨道、乘客使用方式、部分设备系统配置等,另一方面也与常规城市轨道交通有较多的差异性,如票制、运营管理、可采用上车售票,车站内部可采用开敞使用方式、不设付费区及非付费区等。设备可采用外置形式就地设置。由于采用非专用路权及混合路权的形式,车站与区间均可采用半封闭式的管理。和常规城市轨道交通多数采用地下设站不同,有轨电车绝大多数采用地面设站的方式,部分设备配套如充电站也会设置于地面,因此在外观及使用上,有轨电车更多接近于城市常规公交。

由于采用地上设置方式,有轨电车的建设周期比常规城市轨道交通要短,建设投资也相对较小。而采用非专用路权及混合路权的形式常常导致项目建设后,交通效率远远不及地铁,性价比较低。

2. 线路与景观

有轨电车作为新型交通工具的新兴事物出现,无论从交通技术还是运营模式上都是前所未有的探索,也直接影响到车站建筑的设置模式与设计理念,可从功能、文化、历史、景观等方面考虑各种可能性。由于广州海珠环岛新型有轨电车试验段的车站设于临江绿化带中,拥有优美的江北一线景观,在如此优美而休闲的环境中,车站造型的景观化处理能使其成为江边一道靓丽的风景,所以车站在满足必要交通功能的前提下,景观造型设计需要提升到一个更高的层面。有轨电车有别于地铁,地铁系统多设于地下,对城市影响相对较少,有轨电车线路设于城市地面,和城市的关系非常密切。创新是基于对城市环境与历史的尊重,是有轨电车项目最重要的课题,而最典型的是琶醍站。琶醍原为珠江啤酒厂,现已改造为啤酒主题休闲文化餐饮区。线路经过琶醍,需要对琶醍进行较大规模的改造并设一座车站。基于对工业历史文化的尊重,通过对其中一水泵房的旧工业建筑进行改造并成为琶醍站,成为典型的交通文化与工业遗产合一的新型交通建筑。

由于有轨电车采用地面设置的方式,车站体量相对较小,车站形态必然会受到城市规划及用地的相关制约。在南方的气候环境条件下,车站多采用开敞设置的方式,客观上有轨电车沿线景观会形成常规"小品"式的城市界面。

广州海珠有轨电车的设计,有意识从开始就要打造标志性的景观线路和城市界面,为乘客创造一种典型的乘车体验,整条线路采用"直接介入"的方式,试与珠江南岸的城市景观绿化相结合,用绿化景观形成线路的特色。客观来说线通车后乘客在绿化中穿行的体验还是基本实现了"人·车·景"合一的初始标,同时在建设期间最大程度避免对城市交通的干扰。而另一方面,出于对运安全的考虑,线路区间除车站外,均采用封闭的管理模式,却人为割裂了市民珠江的亲水界面和内在的关联,破坏了更高层次的城市景观界面,也带来对有电车的负面评价,这是值得反思的。

与之相反的案例有三亚及珠海有轨电车均采取了路中设置的方式,建成后基本持了城市界面的完整性,也适应了有轨电车非专用路权及混合路权的形式,代是建设期间不可避免地对城市交通造成干扰。

3. 取消与保留

常规城市轨道交通,在建设期间多数会采用大拆大建的行为模式,这也是由地车站较大的空间和规模决定的,而且地铁车站大部分空间位于地下,同时在城轨道交通建设期间,由上至下均采用一种较为强势的政策指引。对沿线阻碍工实施的实体,除非有特殊情况,受制于工期要求及投资限制基本都采取取消单清除的方式,很少考虑与对方结合或保留。

有轨电车相对体量与规模较小,严格来说除车站站台外,其他配套部分的设置是具有一定的灵活性的,因此在沿线遇到阻碍工程实施的实体时,多数倾向于取保留或者融合的处理方式。

广州海珠有轨电车在穿越琶醍区域时,尽量保留了原来港口的工业遗产元素,时将轨道设施与商业区域完全融合在一起,采用占一还一的方式,使交通设施商业空间均能实现效益最大化。车站利用原水泵房改造而成,有效延续了历史信息与工业文化基因。

有轨电车设计中,"侧式"与"岛式"车站形式经过了一段漫长的取舍。

侧式车站有利于车站建设工期安排,但对建设投资和车站功能是不利的。车站中间轨道隔开,导致车站的设备配置及运营管理均要一分为二;客观上侧式车规模要大于岛式车站。

岛式车站则好相反,在建设投资、车站规模、车站功能及运营等方面要明显优侧式车站,但对建设工期安排不利。

从安全性角度,当车站位于路中时,采用侧式车站的安全性相比岛式车站要略一些。而当车站位于地块或绿化带中时,安全性的差距则没有那么明显。

人文与商业

与常规城市轨道交通相比，有轨电车属于一种小级别的交通设施，在城市交通网络中属于从属和配角。因此整体上有轨电车的运营没有专项的补贴几乎是不可能的。

有轨电车为支撑运营，设置商业配套是必然的。受制于车站无法像常规城市轨道交通那样提供足够规模的人流效应，从TOD角度进行大规模商业开发，采用综合体的模式，很难形成理想的土地长期受益，来补贴车站建设和运营。

从建设成本看，有轨电车由于运量级别较低，客流规模不高，也相应带来项目建设的性价比偏低的结果。

基于上述现实及征地的限制，目前有轨电车多数只能采用小商业的形态，如车站广告、售卖机等。再者利用部分地面设备房的剩余空间，在后期运营中改造成为商业空间。

直接的商业形态所带来的效益毕竟是有限的。为弥补这种单一商业形态的不足，广州海珠有轨电车进行了"文化植入"，利用线路经过的站点打造"景观旅游线路"，同时也在车站周边空地、设备站屋面、车辆段等处，利用多种灵活形式及场地安排，以文化活动或临时设施的形式，开展多种准商业形态的造势宣传，力图借助文化活动的平台，催生另类的人流效应和客流模式。也试图通过长期的文化植入，创造特殊的有轨电车商业文化。

所以，有轨电车进一步发展不应局限在交通功能层面，更应充分挖掘其附加值，与城市整合产生更为密切的互动与联系，形成新型的产业价值、文化价值和商业价值，尤其车站功能上可开发出更多便民服务设施和小型创新性试验，使有轨电车与时代进一步结合，以人为本，创造出更多精彩的可能。

The Way of Innovation
(Guangzhou Haizhu Roundabout Tram Test Section)

1. Difference and Unity

As for the transportation positioning of the tram, it is the complementary transportation system for the urban metro system restricted by the demand for transportation capacity and exclusive right-of-way. It is more suitable for the internal transportation connection between the small area and new urban area connected to the urban metro network, realizing the connected radiant network of "small rings linking to big rings" for urban rail transit. As the small-scale rail transit facility, tram shares a lot of similarities with conventional urban rail transit, including vehicle, railway, passenger usage mode and part of equipment system configuration etc., on the other hand, it is also different from conventional urban rail transit in many ways, such as ticket system and operation management, ticket purchase on the tram, the station can adopt open operation method without pay zone and non-pay zone. Equipment can be set in the filed in an external form. Due to the application of non-exclusive right-of-way and mixed right-of-way, the stations and sections can be semi-closed. Unlike most of the conventional urban rail transits to set underground stations, a majority of tramsset the stations on the ground, with part of the ancillary equipment such as charging stations also set on the ground. Therefore the tram is closer to conventional urban public transportation in regard to appearance and utilization.

Owing to the on-the-ground setting, the construction period of the tram is shorter than that of the conventional urban rail transit, and the construction investment is relatively small. The use of non-exclusive right-of-way and mixed right-of-way often leads to lower transportation efficiency and cost performance comparing with metro after the construction.

2. Lines and Landscapes

As a Newly Sprouted Thing of new type vehicle, tram explores in terms of transportation technology and operation mode as never before, and directly affects the setting mode and design concept of station buildings. Various possibilities can be considered in regard of function, culture, history and landscape. Because the station of the Guangzhou Haizhu Roundabout Tram Test Section is located in the riverside green belt and has beautiful waterfront landscape. In such nice and leisure environment, the landscape treatment of the station shape can make it a beautiful landscape line by the river, hence the station design needs to be upgraded to a higher level based on satisfying the necessary transportation functions. Different from the metro which is mostly built under the ground with comparatively less impact on the city, the tram line is set on the ground and closely related to the city. The innovation is the most important topic for the tram program based on the respect for the urban environment and history, among which the most typical one is the Pati Station that used to be the Pearl River Beer factory currently renovated as a beer-themed recreation culture and dining area. When the line was planned to pass through Pati, it required large-scale transform and establishment of a station. Basing on the respect for industrial history and culture, an old industrial building of the pump house was transformed into Pati Station and became a typical new type transportation building with combination of transportation culture and industrial heritage.

Since tram is built on the ground and the station volume is relatively small, the station form is bound to be constrained by urban planning and land use.

Under the climatic conditions of the south, the station usually adopt open setting, objectively the landscape along the tram will form conventional "sketch type" urban interface.

The design of the Guangzhou Haizhu tram has consciously built up an iconic landscape line and urban interface from the beginning to create typical ride experience for passengers. The entire route uses "direct intervention" approach trying to interact with the urban landscape of the south bank of the Pearl River and come up with the characteristic of the line with green landscape. From the objective point of view, passengers' passing through the greening after the line is open basically achieves the initial goal of "passenger, train and landscape" in one, and at the same time avoid interference with urban transportation during construction to the maximum extent. On the other hand, in consideration of operational safety, closed management mode was adopted in the line interval sections except for the station, however it artificially separated the waterside interface and internal connection between the citizens and the Pearl River, spoiled the urban landscape interface of higher-level and brought about negative comments on the tram, which was worth reflecting.

On the contrary, the trams in Sanya and Zhuhai adopts the way of setting on the road, they have basically maintained the integrity of the urban interface and also adapted to the form of non-exclusive right-of-way and mixed right-of-way non-exclusive right-of-way and mixed right-of-way of the tram after completion, and the cost was inevitably interfering the urban transportation during construction.

3. Cancellation and Retention

For conventional urban rail transit, usual practice of large-scale demolition and construction are adopted during the construction period determined basing on the larger space and scale of metro station, moreover, during the construction of urban rail transit, stronger policy guide is applied from top to bottom. Entities that hinder the construction along the line are basically canceled and simply removed as per the requirements of the construction period and investment restrictions unless there are special circumstances, and rarely considered to be combined or retained.

The volume and scale of tram is relatively small, strictly speaking except for station platforms, the settings of other supporting parts are flexible. Therefore it tends to the way of retention or integration when encountering entities that hinder the implementation of projects along the line.

When the Guangzhou HaizhuTram passed through the Pati area, it tried to retain the industrial heritage elements of the original port as much as possible. At the same time the rail facilities were fully integrated with the commercial area to maximize the effectiveness of transportation facilities and commercial space via the take-one-return-one manner. The station transformed from the original pump house effectively extended the historical information and industrial culture genes.

In the design of the tram, the "side-type" and "island-type" station forms had undergone a long trade-off.

The side-type station is good for construction schedule arrangement of the station, but unfavorable for construction investment and station function. Because the station is separated by the track in the middle, the equipment allocation and operation management of the station must be divided into two; objectively the scale of side-type station is larger than island-type station;

The island-type station is the other way around, it is superior to the side-type station in terms of construction investment, station size, station function and operation, but is not good for the construction schedule;

From the safety angle, the side-type station is not as safe as the island-type station. However when the station is located in the middle of the land or the green belt, the safety difference becomes less obvious.

Humanity and Commerce

Compared with conventional urban rail transit, tram is a small-scale transportation facility playing a subordinate and supporting role in urban transportation network. Therefore it is almost impossible for the tram to operate with special subsidies on the whole.

It is inevitable that commercial facilities are deployed for the tram in order to support the operation. Restricted by the fact that the station cannot provide sufficient scale of visitor flow as in conventional urban rail transit, it is hard to generate ideal revenue from the land development through large-scale commercial development from the perspective of TOD using complex model to subsidize the construction and operation of the station.

From the view of construction cost, lower traffic level and passenger flow of the tram also lead to lower cost performance of project construction accordingly. Based on the above-mentioned reality and restrictions on land acquisition, at present most trams can only adopt small business models such as station advertisements, vending machines and so on. In addition, the remaining space of some ground equipment rooms will be used to transform into commercial space in the later operation.

The benefits of direct business model are limited after all. In order to make up for the disadvantage of single business model, the Guangzhou Haizhu Tram implements "cultural implantation", taking advantage of stations on the line to create "landscape tour route", and also using various flexible forms and venue arrangements for cultural activities or temporary facilities in the open space around the station, equipment station roofs and depots, to carry out various business promotion with the effort to facilitate special visit flow effect and passenger flow pattern via cultural activity platform.

It also attempts to create special tram commercial culture through long-term cultural implantation.

Therefore the need for further development of tram should not be limited to the level of transportation function, instead, it should fully exploit its value-added level and create closer interaction and connection with urban integration to form new values of industry, culture and business, especially more convenient service facilities and small innovative tests to be developed for station functions, making tram further join with the times, being people-oriented, and creating more exciting possibilities.

目录

006-009	持续之道（地下车站）
010-013	兼容之道（高架及地面车站）
014-017	文化之道（车站公共区装修）
018-021	多元之道（轨道交通配套建筑及景观）
022-025	创新之道（广州海珠环岛新型有轨电车试验段）
028-103	第一部分　地下车站
104-195	第二部分　高架及地面车站
196-263	第三部分　车站公共区装修设计
264-327	第四部分　轨道交通配套建筑及景观
328-395	第五部分　广州海珠环岛新型有轨电车试验段
396-397	后记
398-399	团队

Content

006–009	The Way of Continuity (Underground Station)
010–013	The Way of Compatibility (Elevated and Ground Station)
014–017	The Way of Culture (Decoration of Station Public Area)
018–021	The Way of Diversification (Metro Auxiliary Building and Landscape)
022–025	The Way of Innovation (Guangzhou Haizhu Roundabout Tram Test Section)
028–103	Underground Station
104–195	Elevated and Ground Station
196–263	Decoration Design of Station Public Area
264–327	Metro Auxiliary Building and Landscape
328–395	Guangzhou Haizhu Roundabout Tram Test Section
396–397	Postscript
398–399	Team

032–047	广州市轨道交通五号线 动物园站	
	Guangzhou Metro Line 5 Zoo Station	
048–049	广州市轨道交通三号线 广州塔站	
	Guangzhou Metro Line 3 Canton Tower Station	
050–051	广州市轨道交通三号线 机场南站	
	Guangzhou Metro Line 3 Airport South Station	
052–053	广州市轨道交通一号线 广州东站	
	Guangzhou Metro Line 1 Guangzhou East Railway Station	
054–055	广州市轨道交通二号线 广州火车站	
	Guangzhou Metro Line 2 Guangzhou Railway Station	
056–057	广州市轨道交通二号线 白云公园站	
	Guangzhou Metro Line 2 Baiyun Park Station	
058–059	广州市轨道交通二号线 江夏站	
	Guangzhou Metro Line 2 Jiangxia Station	
060–061	广州市轨道交通二号线 萧岗站	
	Guangzhou Metro Line 2 Xiao-gang Station	
062–063	广州市轨道交通二号线 白云文化广场站	
	Guangzhou Metro Line 2 Baiyun Culture Square Station	
064–065	广州市轨道交通二号线 飞翔公园站	
	Guangzhou Metro Line Feixiang Park Station	
066–067	广州市轨道交通二号线 黄边站	
	Guangzhou Metro Line 2 Huangbian Station	
068–069	广州市轨道交通三号线 汉溪长隆站	
	Guangzhou Metro Line 3 Hanxi Changlong Station	
070–071	广州市轨道交通三号线 机场北站	
	Guangzhou Metro Line 3 Airport North Station	
072–073	广州市轨道交通四号线 大学城北站	
	Guangzhou Metro Line 4 Higher Education Mega Center North Station	
074–075	广州市轨道交通四号线 广隆站	
	Guangzhou Metro Line 4 Guanglong Station	
076–077	广州市轨道交通四号线 飞沙角站	
	Guangzhou Metro Line 4 Feishajiao Station	
078–079	广州市轨道交通四号线 大涌站	
	Guangzhou Metro Line 4 Dachong Station	
080–081	广州市轨道交通九号线 飞鹅岭站	
	Guangzhou Metro Line 9 Fei'eling Station	

第一部分 地下车站
Underground Station

082–083	广州市轨道交通九号线 花都汽车城站
	Guangzhou Metro Line 9 Huadu Autocity Station
084–085	广州市轨道交通广佛线 蠔岗站
	Guangzhou Metro Guangfo Line Leigang Station
086–087	广州市轨道交通广佛线 南海汽车城站
	Guangzhou Metro Guangfo Line Nanhai Autocity Station
088–089	广州市轨道交通二十一号线 凤岗站
	Guangzhou Metro Line 21 Fenggang Station
090–091	广州市轨道交通七号线二期 长洲站
	Guangzhou Metro Line 7 Phase 2 Changzhou Station
092–093	广州市轨道交通十一号线 广州火车站
	Guangzhou Metro Line 11 Guangzhou Railway Station
094–095	广州市轨道交通十一号线 梓元岗站
	Guangzhou Metro Line 11 Ziyuangang Station
096–097	广州市轨道交通十一号线 广园新村站
	Guangzhou Metro Line 11 GuangyuanXincun Station
098–099	广州市轨道交通七号线二期 深井站
	Guangzhou Metro Line 7 Phase 2 Shenjing Station
098–099	广州市轨道交通七号线二期 洪圣沙站
	Guangzhou Metro Line 7 Phase 2 Hongshengsha Station
098–099	广州市轨道交通十号线 东晓南站
	Guangzhou Metro Line 10 Dongxiaonan Station
100–101	广州市轨道交通七号线二期 石溪站
	Guangzhou Metro Line7 Phase 2 Shixi Station
100–101	广州市轨道交通七号线二期 大干围站
	Guangzhou Metro Line7 Phase 2 Daganwei Station
100–101	广州市轨道交通十号线 东沙站
	Guangzhou Metro Line 10 Dongsha Station
102–103	广州市轨道交通十号线 广钢新城站
	Guangzhou Metro Line 10 Guanggang New Town Station
102–103	广州市轨道交通十号线 西朗站
	Guangzhou Metro Line 10 Xilang Station
102–103	广州市轨道交通八号线北延段 东晓南站、同福西站
	Guangzhou Metro Line 8 North Extension Design of Dongxiaonan Station & Tongfuxi Station

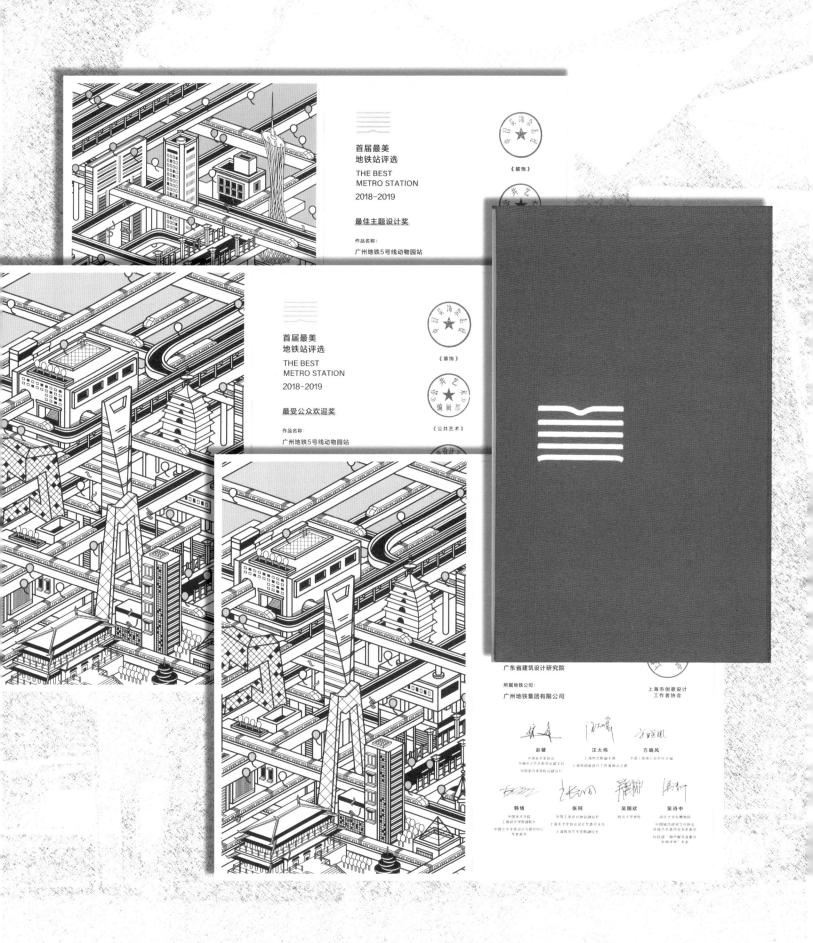

站厅室内实景

广州市轨道交通五号线
动物园站
Guangzhou Metro Line 5
Zoo Station

车站位于广州市环市东路动物园广场前，2009年12月28日开通。
车站具有以下特殊性及创新点：
1. 顺应民意呼声设置车站
2. 工程条件催生特殊站型
线路条件上下叠加暗挖站台，施工条件催生明暗挖分离式车站形式。
3. 车站站型衍生"中庭"站厅
利用深埋三层明挖站厅形成中庭式空间，为乘客提供视线开敞的交通导向，站厅层配合闸机设置改变柱位设置，形成双"Y"形柱结构，成为高大站厅空间的视觉焦点。

4. 地域环境激发设计灵感
动物园的原生态环境及地域特色，激发了设计以清水混凝土作为主题的灵感，水混凝土粗犷、原始的材质特质呼应车站所处地域环境，车站墙面采用大红色瓷钢板为主色配以精心的室内照明设计，进一步突出了动感活力的地域元素。
5. "以少为多"顺应"低碳环保"
车站装修设计采取"以少为多"设计手法，不仅整合了车站设备空间的视觉面，简化了车站装修界面，降低了车站装修造价，有效提高了工程施工效率，顺应了当代"低碳"的设计潮流，提升了车站的空间品质。

地下车站

总平面图

站厅室内实景

站厅室内实景

转换夹层实景

站厅楼梯实景

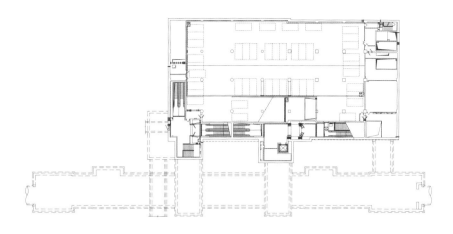

车库（负一）层平面图

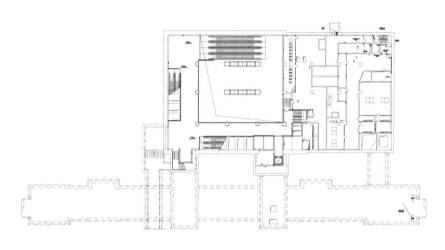

设备（负二）层平面图

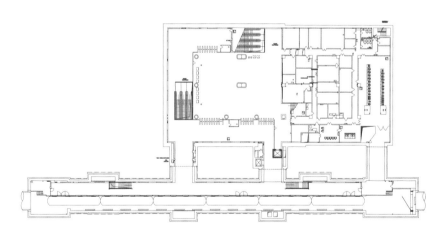

站厅（负三）层平面图

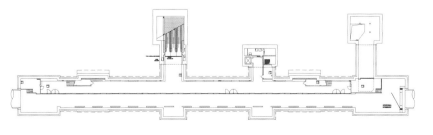

右线站台（负四）层平面图

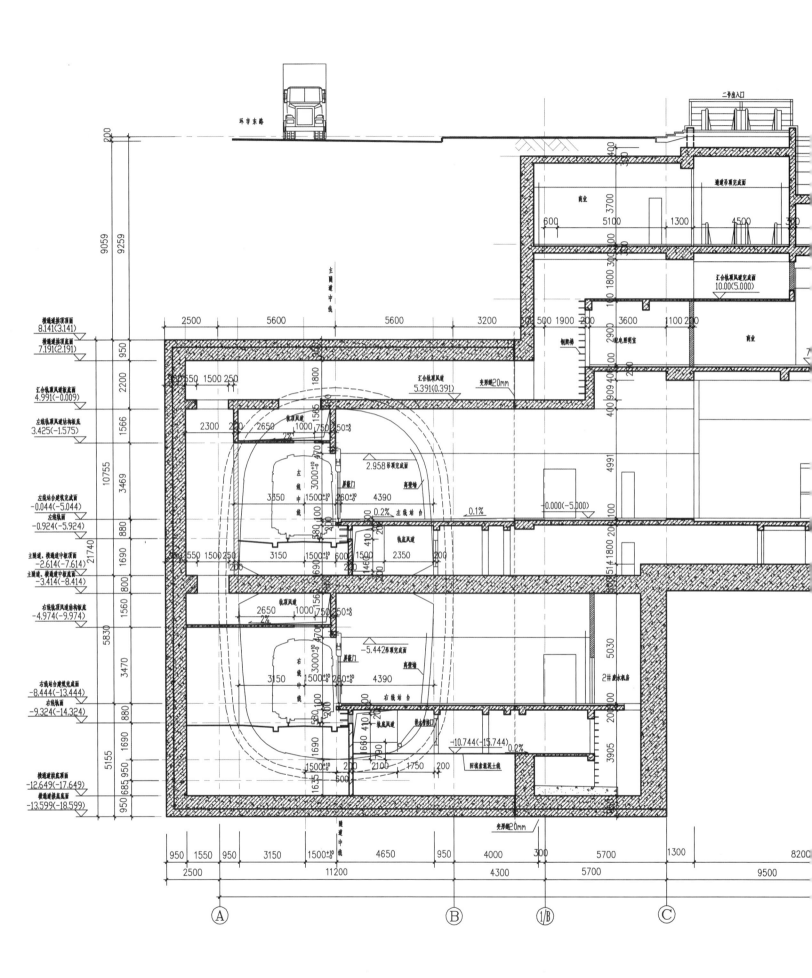

横剖面图

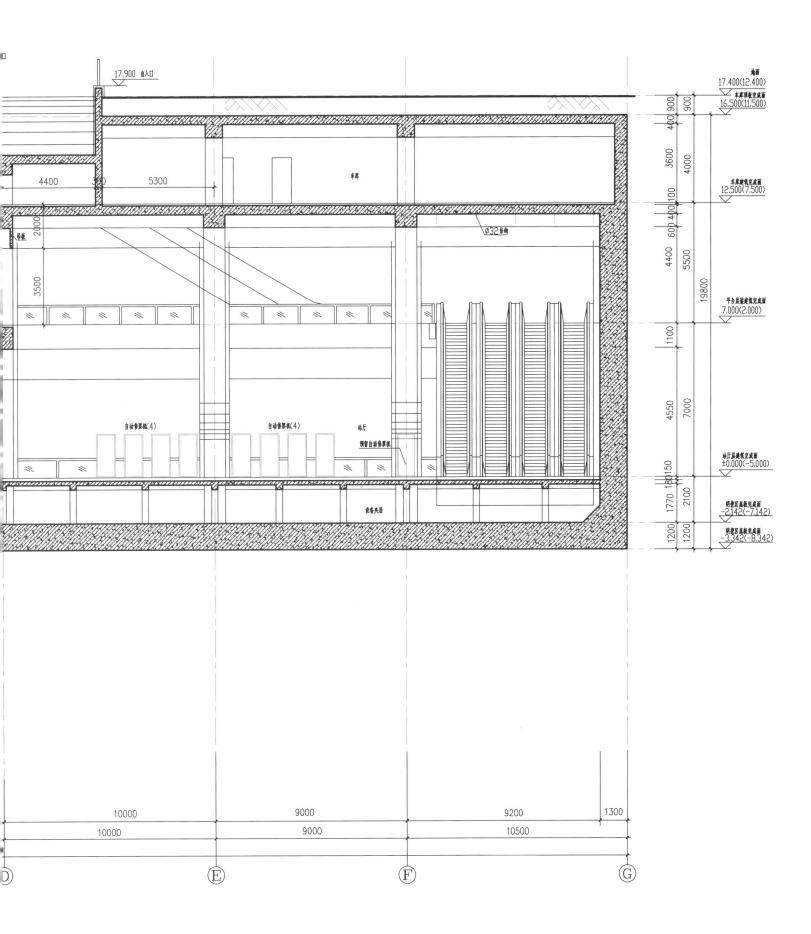

站厅室内实景

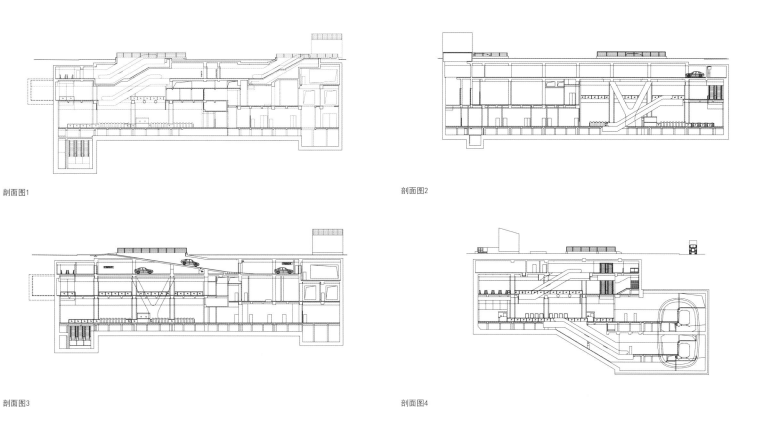

剖面图1

剖面图2

剖面图3

剖面图4

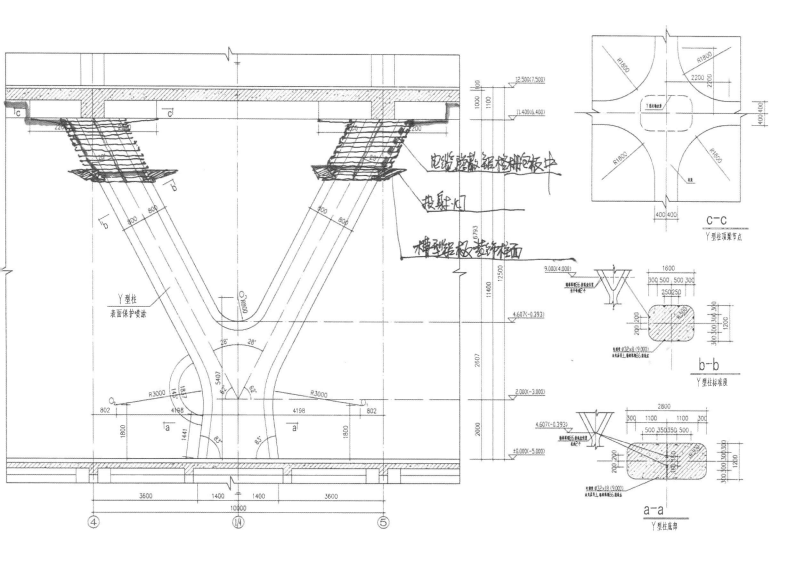

"Y"形柱结构大样图

站厅室内实景

"Y"形柱结构局部实景

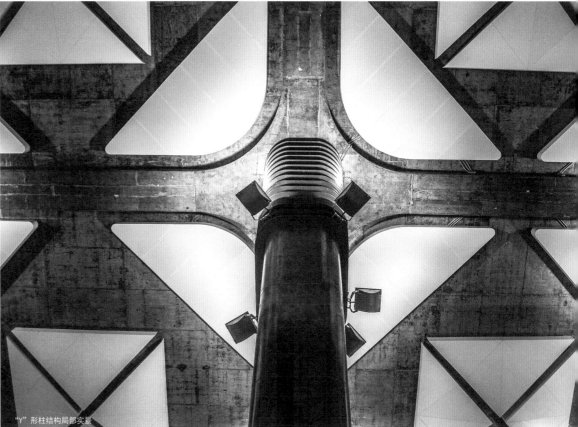

"Y"形柱结构局部实景

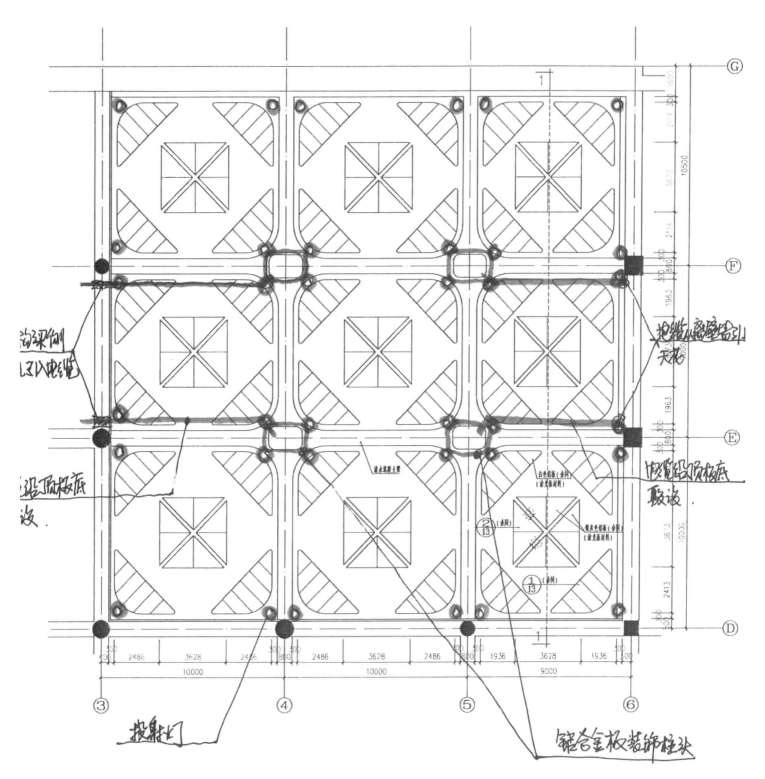

站厅吊顶平面布置图

站厅室内实景

站厅结构梁局部实景

形柱结构局部实景

站台柱内实景

站台换乘楼梯实景

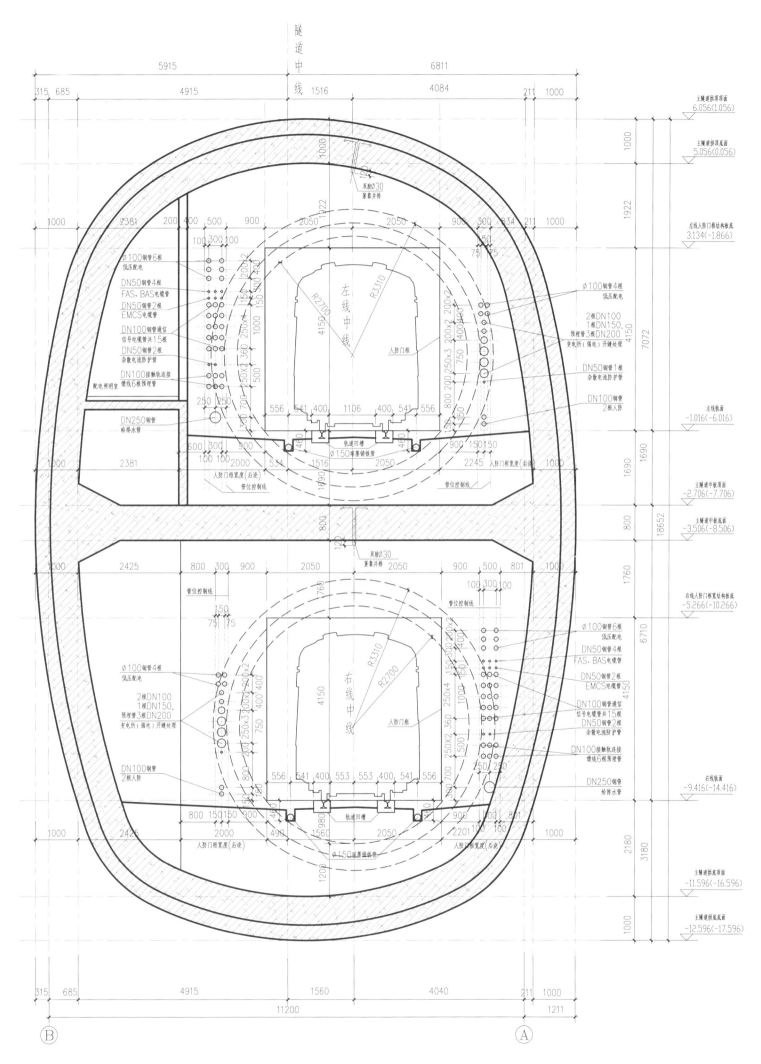

区间剖面图

站厅室内实景

广州市轨道交通三号线
广州塔站
Guangzhou Metro Line 3
Canton Tower Station

广州塔站于2008年开通，车站位于广州市海珠区阅江路，地处珠江江畔，临近广州塔，在突出的广州地标之下设计出更高标准的地铁车站。车站设计利用车站埋深大的特殊条件，打造了广州第一个设有两层中庭式站厅空间的地铁车站，两层通高的中庭式车站打破了传统地下地铁车站低矮压抑的空间特性，为乘客提供了舒适的乘车体验。

车站共3层，车站顶板埋深2m，负一层连接附属出入口，负二层设置票务及付区等功能，负三层为岛式站台及设备区。车站总建筑面积约为11000m²，车站长度148.2m，站台有效长度120m，采用岛式8m站台。车站采用6B编组，共设三个出入口。

地下车站

横剖面图

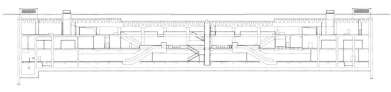
纵剖面图

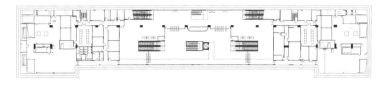
站厅平面图

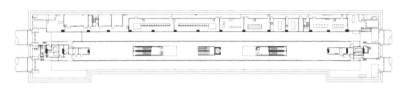
站台平面图

站厅换乘实景

站厅室内实景

总平面

站厅室内实景

广州市轨道交通三号线
机场南站
Guangzhou Metro Line 3
Airport South Station

车站于2010年9月开通，站位于规划中的新白云国际机场T1航站楼停车大楼以及交通中心的地下层，与航站楼同步建设。站厅空间通过中庭与扶梯和机场T1航站楼连接并融为一体。在交通流线设计上考虑与T1航站楼以便捷、无缝的方式紧密连接，车站站厅成了两种重要交通工具之间的转换空间。负一层为站厅层，通过中庭与机场航站楼直接联通，负二层为站台层。

车站总建筑面积11879m²，车站总长度157m，有效长度120m，采用侧式台，线间距5m，站台宽度3m，采用6B编组。车站不设独立出入口，站厅与站楼直接相连接，为航站楼乘客提供换乘。

地下车站

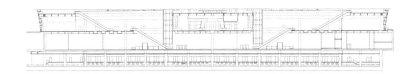
纵剖面图

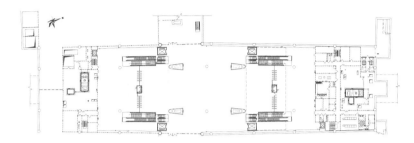
站厅平面图

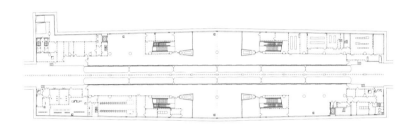
站台平面图

站厅室内实景

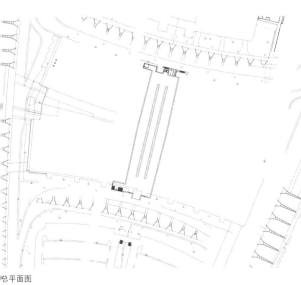
总平面图

站台室内实景

站厅室内实景

广州市轨道交通一号线

广州东站
Guangzhou Metro Line 1
Guangzhou East Railway Station

车站于1998年6月开通，站位位于天河广州东铁路新客运站站房下方，为广州首个与上部大型铁路客站结合设计的枢纽车站。车站重点解决大人流集散、地铁换乘及与城市周边不同设施的衔接，通过站厅放射出多条重要的地下连接通道，连接城市周边公共停车场、公共汽车总站、商业酒店配套等各种配套设施。

车站设有站后折返线。车站总长度560余米，总面积2.2万m^2，为地下二层岛车站。站厅与铁路客站地下一层相连，并与地面公交场地下车库形成立体无接驳的换乘体系，是广州东部大型交通枢纽。

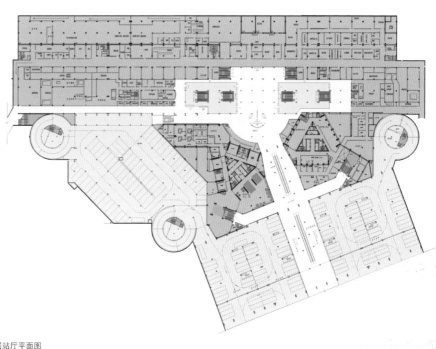

地下负一层站厅平面图

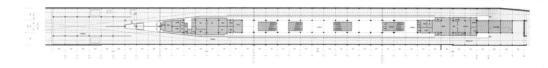

地下负二层站台平面图

站台室内实景

站厅室内实景

站厅室内实景

广州市轨道交通二号线
广州火车站
Guangzhou Metro Line 2
Guangzhou Railway Station

车站位于广州市中心区、人口密度最高的流花火车站广场下方。现状为两线换乘，规划为四线换乘，车站通过扩大的站厅空间重点解决巨大的交通客流集散与组织，集中解决地铁与地铁的换乘及客流集散。车站采用地下二层岛式站台设计，总长147m，总建筑面积1.7万m²。

为有效疏解地面庞大的客流，站厅采用"扩大式"站厅设计，并首次在广州地铁中采用观光式无障碍电梯及过街通道，设置扶梯、地面式设备房等。2002年12月28日车站开通后受到各界好评。

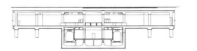
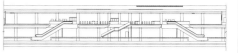

公共区横剖面图　　　　　公共区纵剖面图

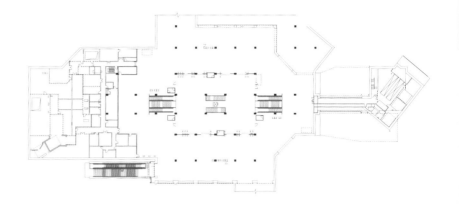

站厅平面图

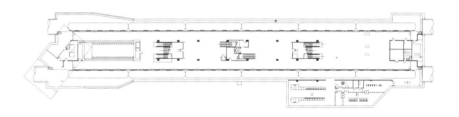

站台平面图

站厅室内实景

站台室内实景

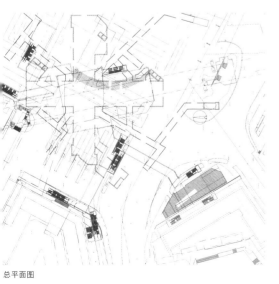

总平面图

地下车站

站厅与站台室内实景

广州市轨道交通二号线
白云公园站
Guangzhou Metro Line 2
Baiyun Park Station

车站于2010年10月开通，位于广州市白云区白云新城核心区，临近白云国际会议中心，受线路高度影响，本站为双层超浅埋侧式车站，站厅与站台同层布置，减少地铁交通空间的压抑，有利于形成良好的交通视线向导，流线上极大提高地铁交通的使用效率。

本站是双层超浅埋车站，顶板覆土最薄处仅约0.5m，二号线车站为单层，站厅站台同层设置，局部负二层预留远期十四号线站台。

车站总建筑面积11222m²，站台总长度198m，有效站台120m，站台宽度3m采用侧式站台，线间距为5m，在站台两端设置了过轨通道。车站共设置了4出入口。

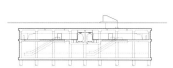

横剖面图

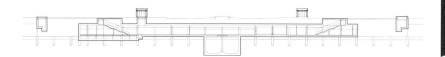

纵剖面图1

纵剖面图2

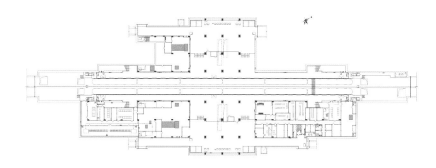

站厅平面图

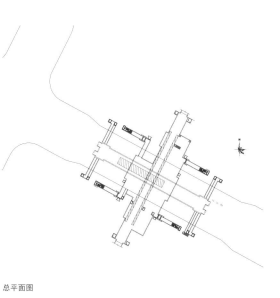

总平面图

站厅与站台室内实景

站厅与站台室内实景

站厅室内实景

广州市轨道交通二号线
江夏站
Guangzhou Metro Line 2
Jiangxia Station

车站于2010年10月开通,位于广州市白云区江夏北二路和黄石北路交界,临近白云尚城住宅区。由于所处区域属岩溶发育较强烈的地段,地质条件较差,车站埋深不宜大。车站为地下一层浅埋侧式,为尽可能减小车站规模,站厅空间紧凑高效布置,站厅与站台同层,形成良好的交通视线向导,提高地铁交通的使用效率。

车站为地下一层浅埋车站,顶板埋深2.5~3.0m,站厅站台同层,设3m宽式站台,分别于站台两端及中部设三组下过轨联系通道。车站总建筑面9328.6m^2。车站总长度174m,站台有效长度120m,线间距4.8m。车站采用编组,设置四个出入口。

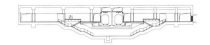

横剖面图

纵剖面图

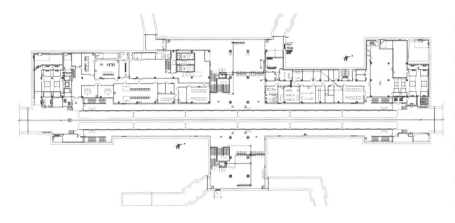

站厅与站台平面图

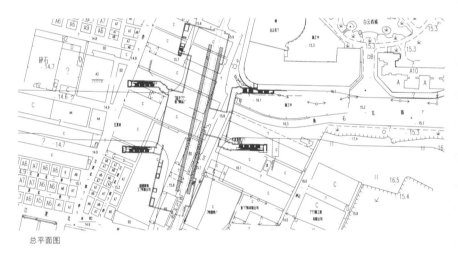

总平面图

站厅室内实景

站台转换楼梯实景

站厅室内实景

广州市轨道交通二号线
萧岗站
Guangzhou Metro Line 2
Xiao–gang Station

车站于2010年8月开通，位于广州市白云区，旧白云机场北端待发展区，并穿越既有的萧岗涌。由于本站是双层超浅埋车站并设有存车线，所以站厅空间也相应加宽并设为双柱站厅，站厅空间较为宽敞明亮。顶板覆土最少处仅约0.5m，设计上考虑本站北端作局部单层处理，仅有站台层，留出站厅层的空间供河涌及管线通过。

车站设一个8m宽岛式站台和一个3m宽侧式站台，中间设一条存车线。车站长为375.50m，总建筑面积18860.0m²。站台有效长度120m，线间距4.8m。站采用6B编组，共设置三个出入口。

地下车站

站台室内实景

站厅室内实景

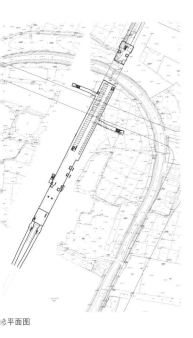

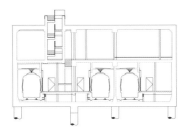
总平面图

横剖面图

站厅平面图

站台平面图

站厅室内实景

广州市轨道交通二号线
白云文化广场站
Guangzhou Metro Line 2
Baiyun Culture Square Station

车站于2010年8月开通，站位位于新规划的白云新城绿化中轴线上，白云国际会议中心西侧的大型绿化广场中央。本站是双层超浅埋车站，站厅与站台同层布置且站厅较为宽敞，减少了地铁交通空间的压抑，有利于形成良好的交通视线向导，流线上极大地提高了地铁交通的使用效率。

本站顶板覆土最少处仅约0.5m，顶板埋深：0.57~1.6m。站厅站台同层设置于地下一层，分设于左右线两侧，设3m宽侧式站台，左右线站台直接与相应站厅连，左右线付费区设置三条地下过轨通道相连。车站总建筑面积9899.3m²，站总长度189.9m，站台有限长度120m，线间距4.8m，车站采用6A编组，设三个出入口。

地下车站

横剖面图

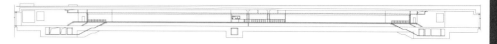
纵剖面图

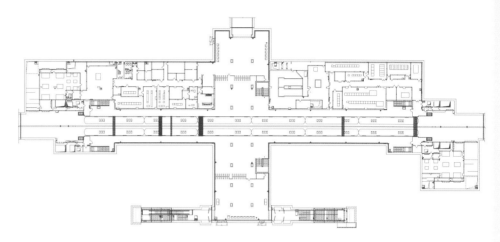
站厅与站台平面图

站台室内实景

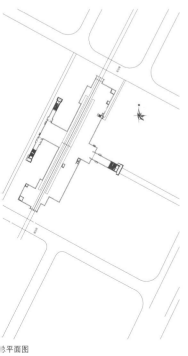
平面图

转换站台楼梯实景

站厅室内实景

广州市轨道交通二号线
飞翔公园站
Guangzhou Metro Line Feixiang Park Station

车站于2010年8月开通，站位位于白云新城绿化中轴线上，地块的东、西以及南侧规划是商贸办公区。为尽可能减小车站规模，站厅空间紧凑高效布置，站厅与站台同层，形成良好的交通视线向导，提高地铁交通的使用效率。
绿地中的车站主体顶板埋深按0.5m控制，南北两端穿规划道路处按1.5～2m控制。

站厅站台同层设置于地下一层，分设于左右线两侧，付费区设置三条地下轨通道相连。车站总建筑面积8828m²，设3m宽标准侧式站台，车站总长212m，有效站台长度140m，线间距4.8m，采用6B编组，共设置两个出入口

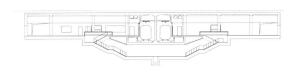

横剖面图

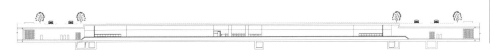

纵剖面图

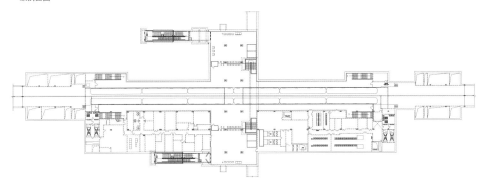

站厅与站台平面图

站厅室内实景

转换站台楼梯实景

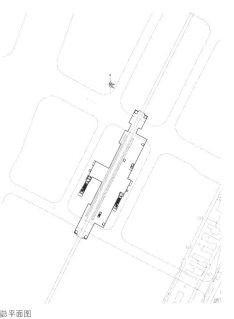

总平面图

地下车站

站厅室内实景

广州市轨道交通二号线
黄边站
Guangzhou Metro Line 2
Huangbian Station

车站于2010年8月开通，位于广州市白云区白云新城，线路受岩溶等不良地质条件影响。考虑浅埋布置，本站设计为地下明挖单层车站，覆土3m。站厅与站台同层布置，减少地铁交通空间的压抑，有利于形成良好的交通视线向导，流线上极大提高地铁交通的使用效率。

车站设置同层双站厅、侧式站台站（车站两端以及中部下过轨换乘），分设左右线两侧。车站总建筑面积约9795.9m²，车站总长度约183.6m，有效站台长度140.00m，站台宽度3.0m，采用6B编组，线间距4.8m，共设置四个出入口

地下车站

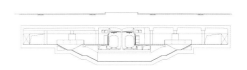

横剖面图

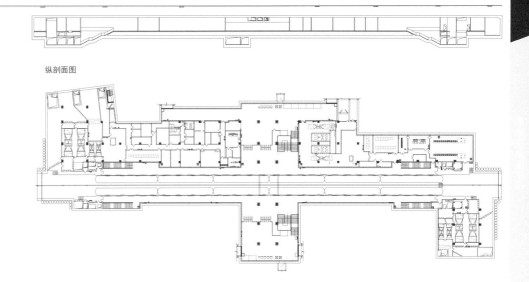

纵剖面图

站厅与站台平面图

站厅室内实景

站台室内实景

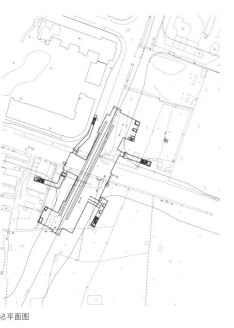

总平面图

站厅室内实景

广州市轨道交通三号线
汉溪长隆站
Guangzhou Metro Line 3
Hanxi Changlong Station

车站于2006年12月开通，位于广州市番禺区万博中心与番禺长隆公园区附近，车站的站厅站台为标准的单柱空间，以车站客流引导为核心展开一个有序列性的交通空间。

车站采用两段设备房的布置形式，为节约车站长度，部分电气设备房布置于站厅旁，靠近负荷中心，节省了管线敷设的距离。

车站顶板覆土2m，车站2层，负一层站厅，负二层站台，采用岛式站台，站台宽度10m。总建筑面积约12400m²，车站长度175m，站台有效长度120m，采用6B编组，共设置三个出入口。

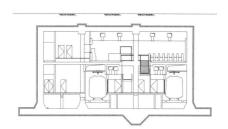

横剖面图

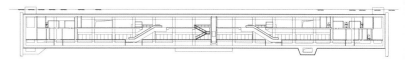

纵剖面图

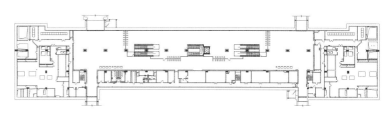

站厅平面图

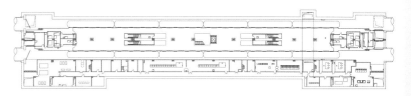

站台平面图

站厅室内实景

站台室内实景

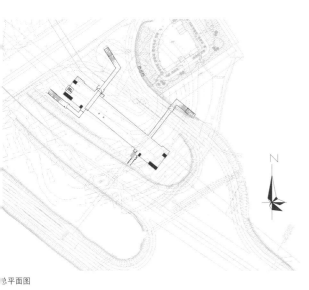

总平面图

站厅室内实景

广州市轨道交通三号线
机场北站
Guangzhou Metro Line 3
Airport North Station

车站于2018年开通,为三号线北延段终点站。站位位于规划中的新白云国际机场T2航站楼停车大楼以及交通中心的地下层,与白云机场T2航站楼同步建设。车站为与机场T2航站楼连接的换乘枢纽站,为方便客流换乘,站厅结合换乘流线设计为多柱宽敞式空间,层高局部增高,满足大客流对空间使用的要求。

环控机房与管理用房主要设置于站厅层,站台层设置变电所。变电所夹层位于变电所下方,通过局部挖深处理。新航站楼通过交通中心与停车大楼连接,机场客通过交通中心的地面楼扶梯与地铁车站换乘。

总建筑面积约19445m²,车站埋深约17.523m,总长约262.5m,有效站台长120m,总宽约63.8m,车站采用侧式站台,站台宽度3m,采用6B编组,车站后为折返线。

地下车站

站厅室内实景

站台室内实景

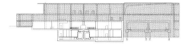
横剖面图

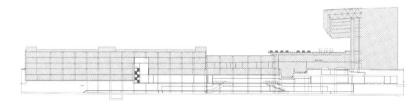
纵剖面图

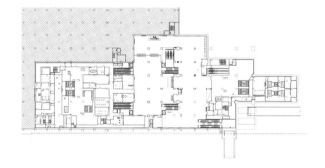
站厅平面图

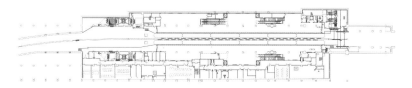
站台平面图

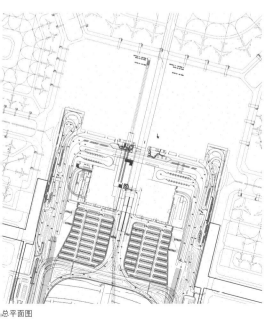
总平面图

071

站厅室内实景

广州市轨道交通四号线
大学城北站
Guangzhou Metro Line 4
Higher Education Mega Center North Station

车站于2005年12月开通，位于广州市番禺区大学城中心北大街，工程与广州大学城的规划、建设同期进行。结合道路布置，车站风亭以低矮敞口形式设置在中央绿化带上，可有效降低对周边景观环境的影响。本站是广州地铁中首次采用存车线站内设置，为有效利用本站规模较大的特点，站内设置集中冷站，为前后4个车站供冷，其他站不设置制冷机房及冷却塔，可减少规模及对环境的影响。车站采用4L编组，总建筑面积13922m^2，为一岛一侧站台三线车站。车站总长323.6m，有效站台长度72m，标准段宽26.2m。

地下车站

站厅室内实景

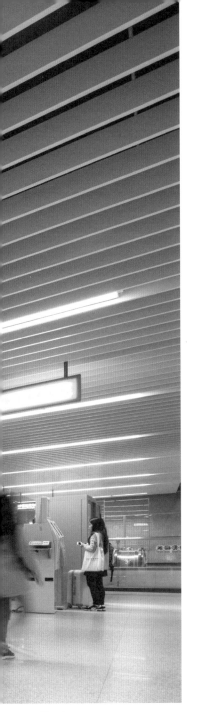

站台室内实景

站厅局部实景

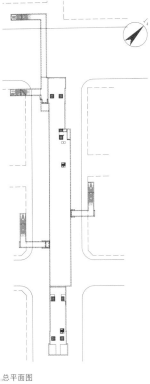

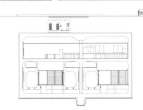
横剖面图

纵剖面图

站厅平面图

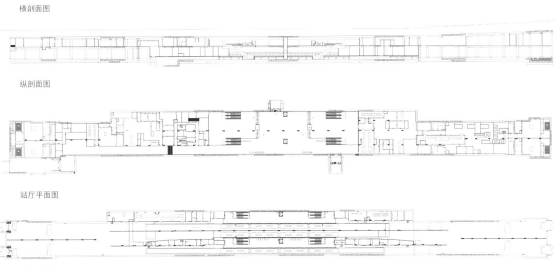
站台平面图

总平面图

站厅室内实景

广州市轨道交通四号线
广隆站
Guangzhou Metro Line 4
Guanglong Station

车站于2017年12月开通，位于南沙区环市大道与广兴路（海滨路）交叉路口沿环市大道布置，在站台西端设有出入场线接南沙停车场。车站的站厅站台为标准的单柱空间，以车站客流引导为核心展开一个有序列性的交通空间组织。总建筑面积27118.92m²。车站为地下2层，负一层为站厅，负二层为站台，为岛式站台车站，全长506.0m，标准段宽19.8m。采用岛式站台，有效站台长72m，宽度11m。车站采用4L编组，共设10个出入口（含预留物业开发空间和两组风亭、一个下沉式冷却塔放置井，位于物业区的出入口暂不实施。

地下车站

站厅室内实景

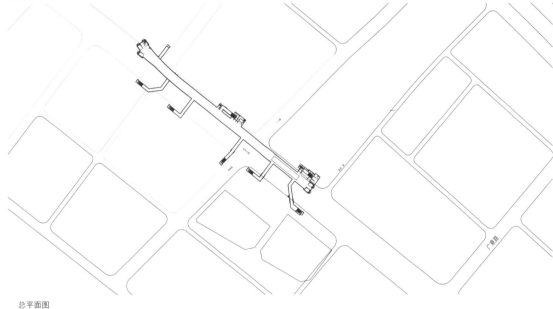

总平面图

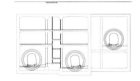

横剖面图

纵剖面图

站厅平面图

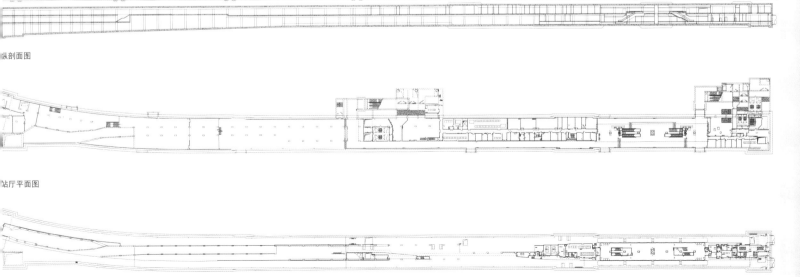

站台平面图

站厅室内实景

广州市轨道交通四号线
飞沙角站
Guangzhou Metro Line 4
Feishajiao Station

车站于2017年12月开通，位于南沙区金隆路西侧地下，临近南沙停车场。车站的站厅站台为标准的单柱空间，通过车站两组楼扶梯及一组垂直电梯以车站客流引导为核心，展开一个有序列性的交通空间组织。

车站总建筑面积9596m²，车站顶板覆土3.5m。车站为地下2层，岛式站台车站，负一层为站厅，负二层为站台，外包总长200.8m，标准段外包总宽20m。有效站台长度72m，宽度11m，采用4L编组。车站设置3个出入口，风亭及冷却塔均布置车站顶板上方，位于南沙停车场用地范围内。

地下车站

楼梯局部实景

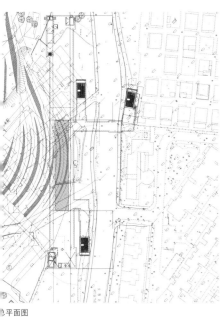

总平面图

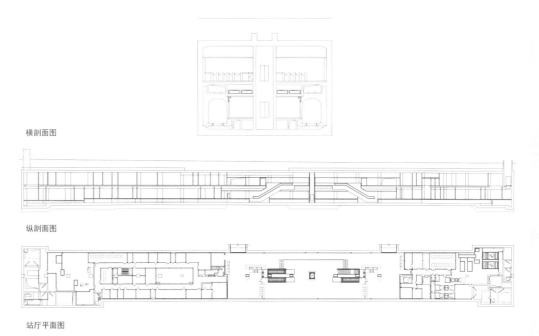

横剖面图

纵剖面图

站厅平面图

站厅室内实景

广州市轨道交通四号线
大涌站
Guangzhou Metro Line 4
Dachong Station

车站于2017年12月开通，位于南沙区环市大道与工业四路（大涌路）交叉路口东侧沿环市大道布置。车站的站厅站台为标准的单柱空间，通过车站两组楼扶梯及一组垂直电梯作为交通核心，为车站展开一个有序列性的交通空间组织向导。
车站采用4L编组，总建筑面积11980.94m²。车站为地下2层，负一层为站厅，负二层为站台；岛式站台车站，全长193.0m，标准段宽19.8m，有效站台长72m，宽度11m，车站采用4L编组。共设三个出入口，两个外挂风亭，一个沉式冷却塔放置井。

地下车站

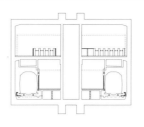

横剖面图

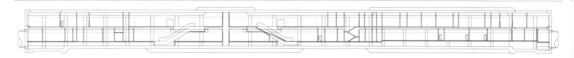

纵剖面图

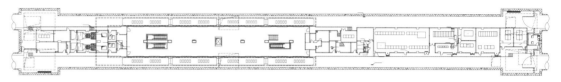

站台平面图

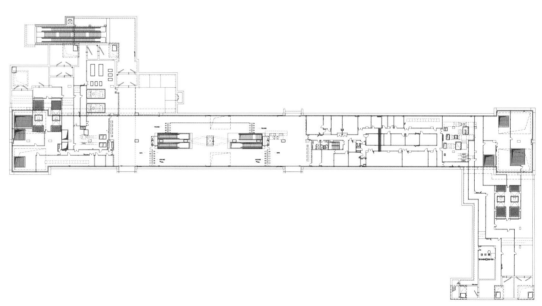

站厅平面图

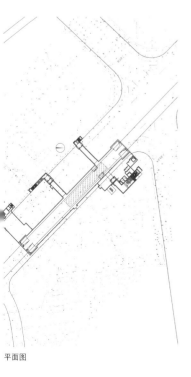

平面图

站厅室内实景

站厅室内实景

广州市轨道交通九号线
飞鹅岭站
Guangzhou Metro Line 9
Fei'eling Station

车站于2017年12月开通。飞鹅岭站设于风神大道西段，飞鹅岭附近，线路沿风神大道行进，在该段走向基本为东西向。车站为标准单柱车站，车站强化交通功能核心区，站厅站台的中部吊顶与柱子结合，形成线路花都之"红花"的主题，形成视觉焦点，强化交通核心功能。

本站位为九号线起点站，西侧接车辆段。车站总建筑面积20550.4m²，为地下2层岛式站台车站，线间距13m，站前设交叉渡线，站后设单渡线。车站总长381.50m，标准段宽19.90m。

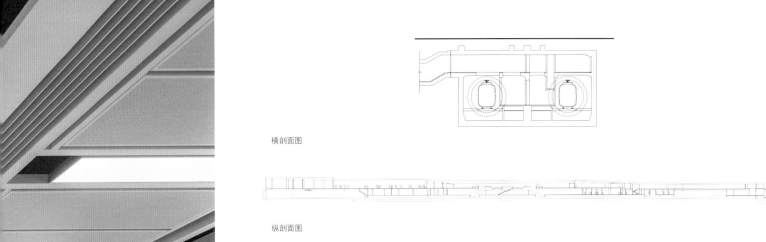

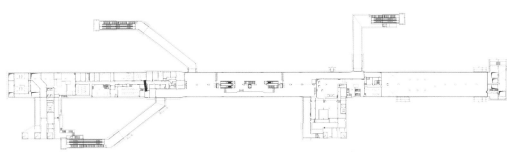

横剖面图

纵剖面图

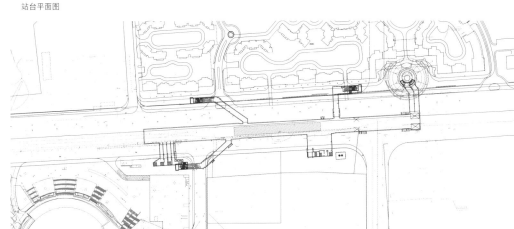

站厅平面图

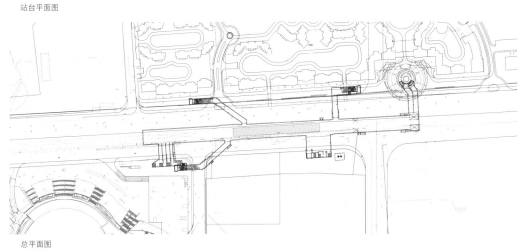

站台平面图

总平面图

站厅室内实景

站台室内实景

站厅室内实景

广州市轨道交通九号线
花都汽车城站
Guangzhou Metro Line 9
Huadu Autocity Station

车站于2017年12月开通。花都汽车城站位于风神大道与九潭路的交叉路口，沿城市主干道风神大道呈东西走向。车站为标准单柱车站，车站以强化交通功能为核心区，站厅站台的中部天花与柱子结合，形成线路花都之"蓝花"的主题，形成视觉焦点，强化交通核心功能。

站位所处的风神大道现状及远期规划均为70m宽，南北向道路九潭路现为15m宽，远期规划为30m宽。车站主体建筑面积12275.5m²，车站总长218.5m；车站为地下2层。采用岛式站，线间距13m，标准段宽度19.90m。

地下车站

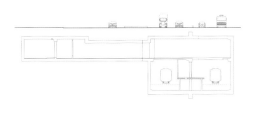

横剖面图

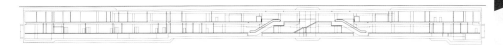

纵剖面图

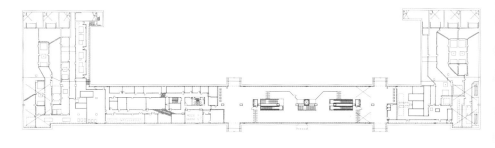

站厅平面图

站台平面图

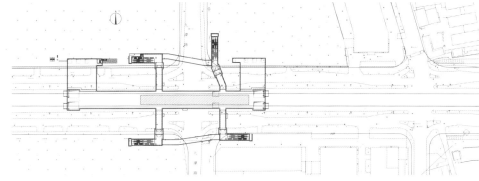

总平面图

站厅室内实景

站台室内实景

站厅室内实景

广州市轨道交通广佛线
蠡岗站
Guangzhou Metro Guangfo Line
Leigang Station

车站于2009年开通。站位位于佛山桂澜路与夏平西路交接处，车站南北走向，为标准的单柱车站，站厅站台空间通过统一的土红色陶土板作为墙面与柱面的装饰材料，突出车站佛山"陶"文化的地域特色，也是全线统一形象的重要特征，强化了线路的识别性。

车站总建筑面积13217m², 车站顶板埋深3m, 车站为地下2层, 负一层站厅负二层站台。车站采用岛式站台, 线间距为13m, 站台宽度10m, 车站总长223m, 站台有效长度为80m, 采用6B编组, 共设置5个出入口。

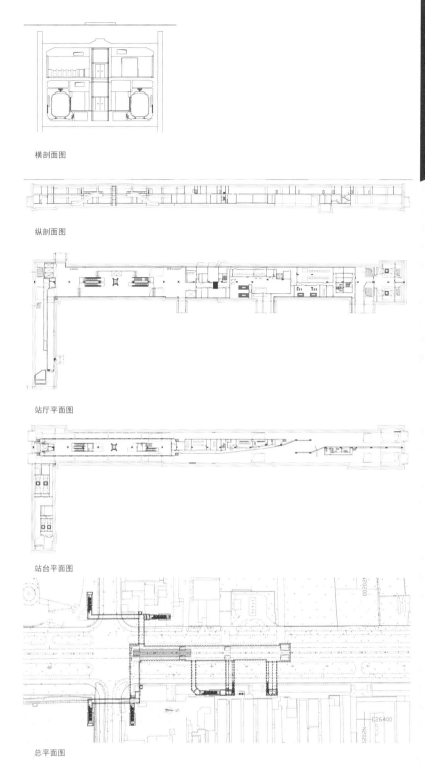

横剖面图

纵剖面图

站厅平面图

站台平面图

总平面图

站台室内实景

站厅换乘局部实景

站厅室内实景

广州市轨道交通广佛线
南海汽车城站
Guangzhou Metro Guangfo Line
Nanhai Autocity Station

南海汽车城站位于南海市桂澜路与海八路（规划81m道路红线）交叉口东南侧规划南海长途汽车站用地内，西端盾构吊出，站后设存车线以及出入段线。车站的站厅站台空间较标准站更为宽敞，所以设计为双柱车站；站厅站台空间通过统一的土红色陶土板作为墙面与柱面的装饰材料，突出车站佛山"陶"文化的地域特色，也是全线统一形象的重要特征，强化了线路的识别性。

南海汽车站采用2层地下车站形式，负一层为站厅层，负二层为站台层，用12m岛式双柱站台，80m有效站台。西端布置车站隧道风机房及部分设备用房，局部扩大设置牵引降压变电所以及其他设备用房。车站总建筑面积9418m²，车站顶板覆土1.25m；车站总长度145m，有效站台长度80m，站台宽度12m。采用XA编组，共设置三个出入口。

地下车站

站台室内实景

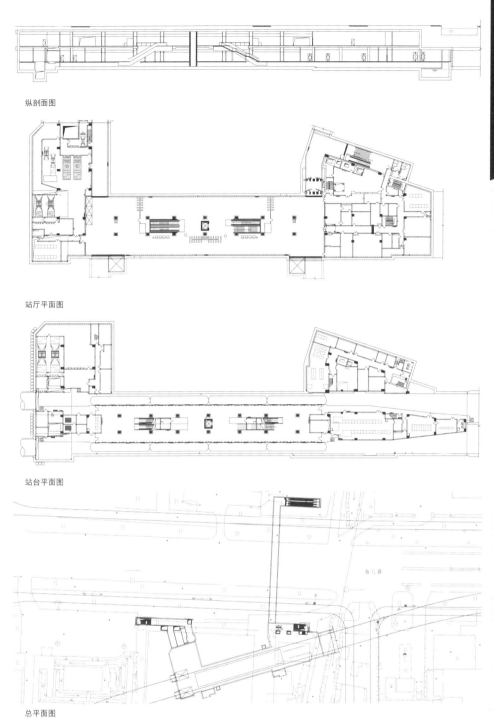

纵剖面图

站厅平面图

站台平面图

总平面图

站台室内实景

站台室内实景

站厅室内实景

广州市轨道交通二十一号线
凤岗站
Guangzhou Metro Line 21
Fenggang Station

车站于2018年12月开通,站位于广州朱村镇广汕路。车站东西走向,为标准的单柱车站,站厅站台的中区通过强烈的波浪形吊顶与柱子的结合,既强化车站交通核心区空间,也为车站展开一个有序列性的交通空间组织向导。

总建筑面积14548m²,车站顶板覆土厚度为3m。车站为地下2层,负一层站厅,负二层站台。车站采用岛式站台,线间距为15m,站台宽度10m,车站长度268m,有效站台长度120m,采用6A编组。共设置4个出入口。

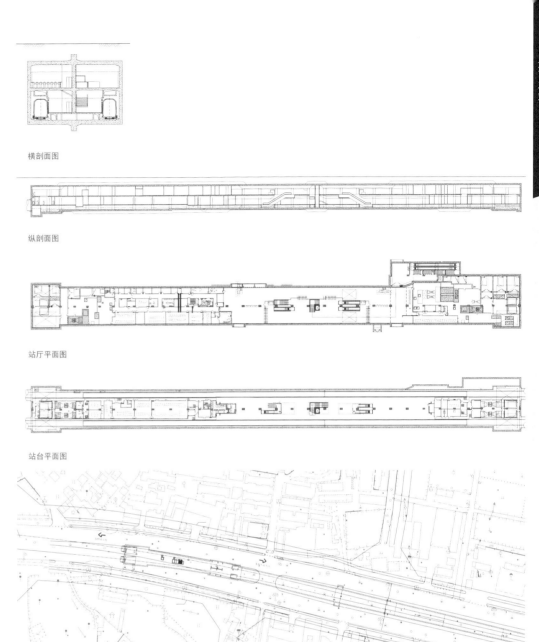

横剖面图

纵剖面图

站厅平面图

站台平面图

总平面图

站厅室内实景　　　　　　　　　站厅室内实景

地下车站

站厅效果图

广州市轨道交通七号线二期
长洲站
Guangzhou Metro Line 7 Phase 2
Changzhou Station

长洲站为七号线二期第三个车站，位于广州市黄埔区长洲岛内，其中七号线车站位于金洲北路与金蝶路交汇处，毗邻广州黄埔军校。车站在设计上考虑呼应黄埔军校的特殊历史文化背景，在车站空间主题设计上考虑表达仰望深远星空、缅怀革命前辈的历史情怀，在引导交通空间同时考虑文化主题的塑造。车站为地下2层岛式车站，设计总长度258.6m，站台宽14m，标准段宽22.70m，有效站台长120m。

车站总建筑面积17869.26m²，主体建筑面积12240.43m²，附属建筑面积5628.83m²。车站顶板覆土3.0m。车站设4个车站出入口（A出入口为远期建），负三层夹层为换乘层。乘客通过换乘层到达八号线站台层，与八号线站"T"形换乘。

地下车站

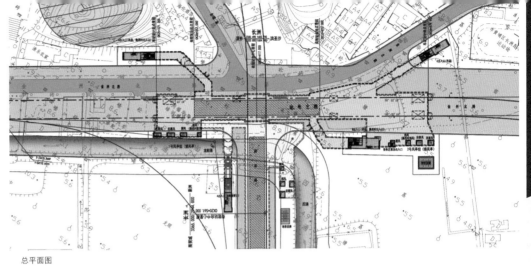
总平面图

站台效果图

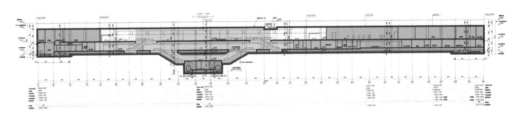
纵剖面图

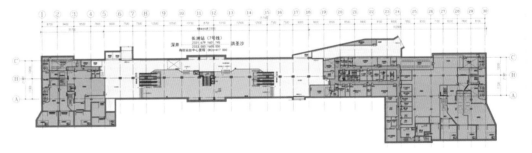
站厅平面图

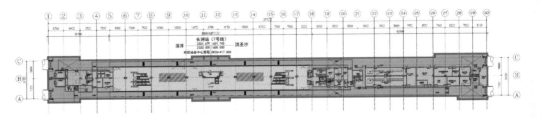
站台平面图

站厅室内实景

广州市轨道交通十一号线
广州火车站
Guangzhou Metro Line 11
Guangzhou Railway Station

广州火车站车站计划于2020年通车。车站位于广州火车站站前广场及现状客运大楼下方，大致呈南北走向，与广州市轨道交通线网既有的二号线、五号线及规划十四号线换乘。四线换乘且受大铁改造的制约，车站需通过对多线复杂流线进行设计研究，重点解决巨大的交通客流集散与组织，集中解决地铁与地铁的换乘和地铁与大铁的换乘方式。站厅层大部分为公共区，地铁设备管理用房设置在北端（国铁站房下方）。

付费区内布置四组楼扶梯及一部电梯，电梯直达站台层。非付费区除设置入口通往地面广场或既有过街通道外，拟布置楼扶梯及电梯直达国铁站房厅。本层公共区面积约8700m^2。车站主体建筑面积40782.93m^2，总建筑面积44804.05m^2。车站总长278.035m，标准段宽38.9m，有效站台宽28.0m，线间距31.2m，站台长度186.0m，共设5个出入口。

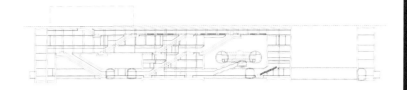
纵剖面图

站厅平面图

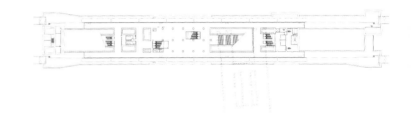
站台平面图

总平面图

台室内实景

车站剖透视效果图

站厅室内实景

广州市轨道交通十一号线
梓元岗站
Guangzhou Metro Line 11
Ziyuangang Station

梓元岗站为广州市轨道交通十一号线第十四个车站，北接广园新村站，南联广州火车站。车站采用分站厅方式设计，车站设计重点需清晰引导分站厅人流的客流组织，避免因引导不清晰导致交通客流交叉，根据暗挖式的结构空间打造现代交通的简捷车站模式。梓元岗站周边道路三元里大道、机场路及解放北路规划路宽均为60m，站位西侧为梓元岗绿地广场，东侧为金桂园小区、花季小区、广州白云皮具城及广东中医药大学。

方案采用明暗挖结合工法，站台暗挖，东西两侧站厅明挖；暗挖站台长153m；线间距35m；主体建筑面积27076m²，总建筑面积30585.8m²。车站设5个出入口，2组风亭，1座下沉式冷却塔；负四层有效站台长度186m，线距35m，设置屏蔽门系统，有效站台右线中心线轨面标高为−16.636m（广州程）。站台层中部为公共区，两端为地铁设备用房。

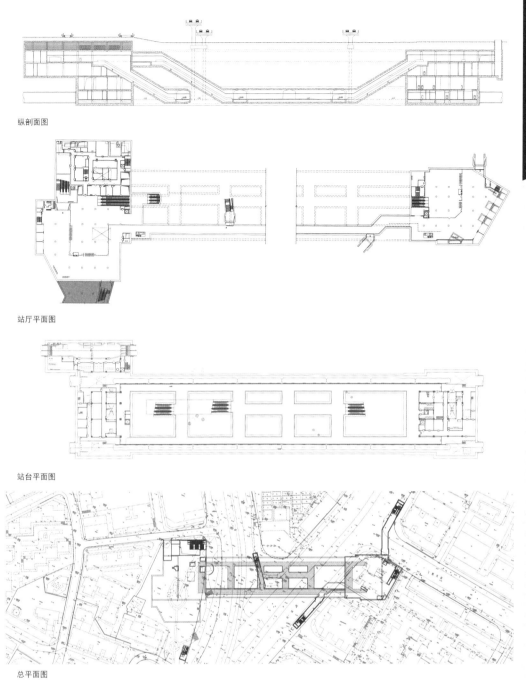

纵剖面图

站厅平面图

站台平面图

总平面图

站台室内实景

地下车站

站厅室内效果图

广州市轨道交通十一号线
广园新村站
Guangzhou Metro Line 11
Guangyuan Xincun Station

广园新村站位于广州市白云区广州中医药大学校内东北角,车站北侧为广园中路,东侧为下塘西路,车站上方为新规划的学校宿舍楼及体育馆。

本站为十一、十二线换乘站,结合周边环境和线路站位实际情况,车站设计成设备区4层,公共区3层的2层通高共用站厅车站,站台为20m的上下叠岛,采用中部换乘梯的异台换乘模式。

本站为广州第一个同时实施的上下叠岛异台换乘站,也是第一个新建车站与地块建筑共同结合设计的换乘站,结合广州地铁未来发展指导纲要,本站拟作为广州地铁的第一个落地试点站。车站设计上创新地将站厅设为两层通高无柱站厅,中部整合设计成中部功能筒,筒中为集公共区竖向交通、智能监控、综合信息播告、车站管理于一体的中央枢纽筒。通高无柱设计和中央枢纽筒为智慧增的无感通行、无人值守、24小时运营等提供了空间条件。

地下车站

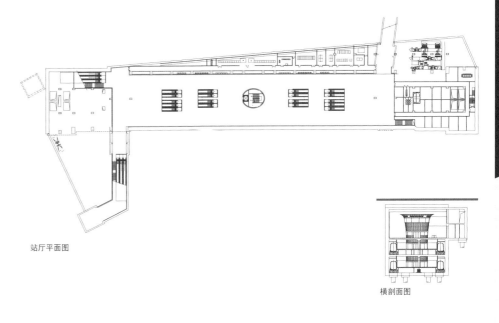

站厅平面图

横剖面图

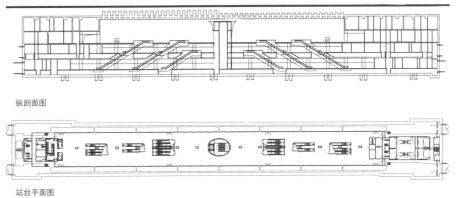

纵剖面图

站台平面图

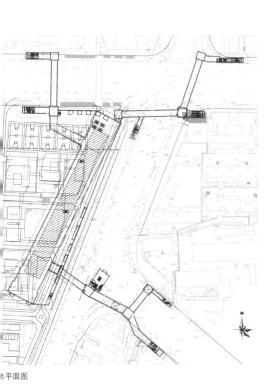

总平面图

站厅室内效果图

站厅室内效果图

097

广州市轨道交通七号线二期
深井站
Guangzhou Metro Line 7 Phase 2
Shenjing Station

深井站位于金洲南路下方，站位所在道路规划金洲南路道路红线宽60m，现状道路宽15m，双向2车道。

深井站为地下两层岛式车站，设计总长度为495m，站台宽13m，标准段宽22.5m，有效站台长120m。现状场地标高8.400m，设计场地标高8.400m，防洪标高8.800m，出入口室内标高9.550m(防洪报告要求)。车站总建筑面积31187.17m²，主体建筑面积22809.91m²，附属建筑面积8377.26m²。车站设4个车站出入口，并与穗莞深城际大学城东站实现L型通道换乘。

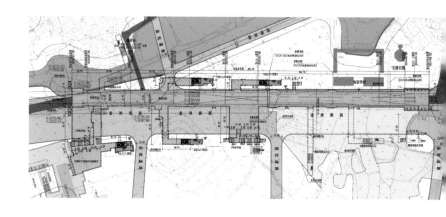

总平面图

广州市轨道交通七号线二期
洪圣沙站
Guangzhou Metro Line 7 Phase 2
Hongshengsha Station

洪圣沙站位于洪圣沙岛西部，设于规划路路中绿化带下方。线路需两次下穿珠江，故出洪圣沙站后以最大坡度下压轨面，尽量让区间位于稳定土层中，降低盾构过江风险，因此车站埋深较大，设置3层车站，顶板覆土约为3.0m。

本站为地下3层12m宽无柱岛式站台车站，设计总长度为172.8m，站台宽12m，标准段宽21.5m，有效站台长度120m，车站总建筑面积16016.25m²，主体建筑面积14247.54m²，附属建筑面积1768.71m²。车站共设置4个出入口。

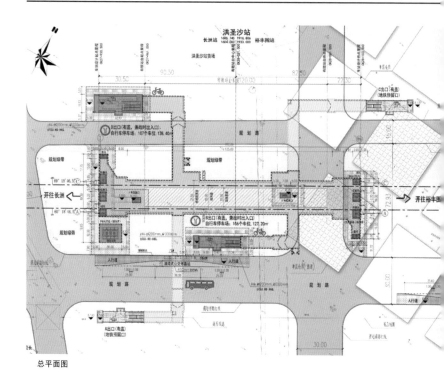

总平面图

广州市轨道交通十号线
东晓南站
Guangzhou Metro Line 10
Dongxiaonan Station

车站位于侨港路与东晓南路交叉口以西，沿侨港路东西向敷设，与既有二号线换乘。车站设置于侨港路，避免横跨东晓南路，减少对周边住宅影响，减少拆迁面积，通过换乘通道与2号线东晓南站进行换乘。车站共设置2个出入口，1个安全疏散口，2个风亭组，1个冷却塔。设计总长度165.5m。站台宽13m，标准段宽28.5m，有效站台长度120m。车站总建筑面积15643.83m²。换乘方式采取通道换乘，十号线东部在负一层设置换乘通道，实现与二号线的付费区和非付费区换乘。

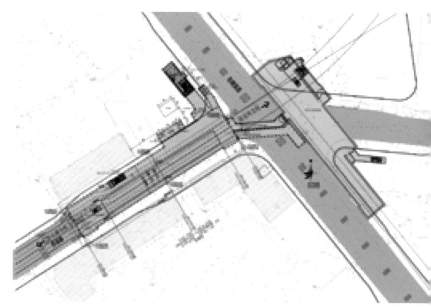

总平面图

地下车站

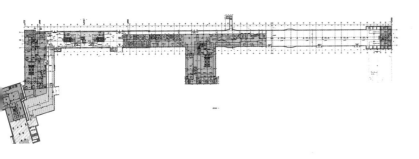
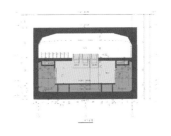

站厅平面图　　　　　　　　　　　　　　　　　横剖面图

站台平面图　　　　　　　　　　　　　　　　　纵剖面图

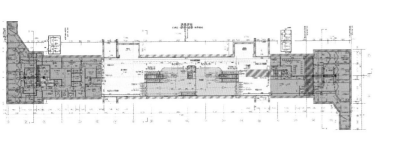
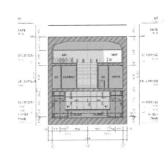

站厅平面图　　　　　　　　　　　　　　　　　横剖面图

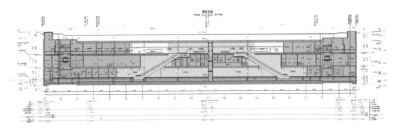

站台平面图　　　　　　　　　　　　　　　　　纵剖面图

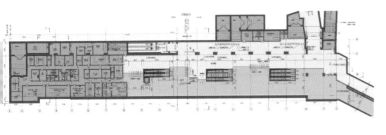

站厅平面图　　　　　　　　　　　　　　　　　横剖面图

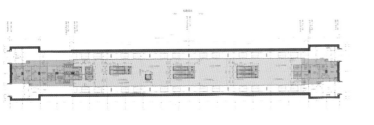
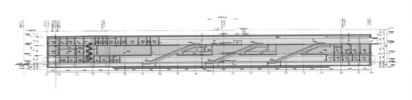

站台平面图　　　　　　　　　　　　　　　　　纵剖面图

广州市轨道交通七号线二期
石溪站
Guangzhou Metro Line 7 Phase 2
Shixi Station

车站位于海珠区工业大道、江南大道与金鹏路的交叉路口处，南联大干围站，北联东晓南站。车站为地下3层（局部4层）岛式车站。设计总长度171.49m，有效站台长120m、宽13m，标准段宽22.7m。车站总建筑面积20317.98m²，主体建筑面积16025.64m²。由于换乘距离较远、换乘客流不大等，车站近期采用AFC票价优惠系统进行非付费区地面换乘，并在十号线站厅层车站侧壁，预留远期付费区换乘接入条件。

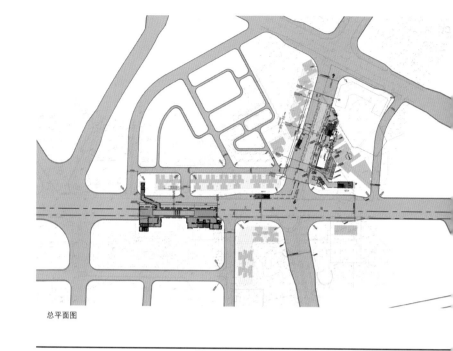

总平面图

广州市轨道交通七号线二期
大干围站
Guangzhou Metro Line 7 Phase 2
Daganwei Station

位于海珠区，站点南侧为珠江，北侧为广州环城高速、石溪等。站点周边现状为工业厂房建筑，地块内有现状河涌。为地下3层岛式无柱车站，共设置4个出入口、2组风亭、1个冷却塔及1个安全出口，D号出入口设置一部无障碍电梯。车站总长度271.7m，标准段宽21.7m，有效站台长120m，宽12m，负一层为站厅层，主体建筑面积5670.5m²，附属建筑面积3920m²。中部为公共区，付费区位于公共区中部，非付费区位于付费区两端。

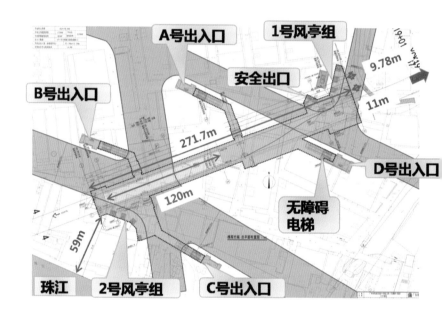

总平面图

广州市轨道交通十号线
东沙站
Guangzhou Metro Line 10
Dongsha Station

本站为广州市轨道交通十号线工程。东沙站位于芳村翠园路上，西侧为东新高速出入口，南侧为广州环城高速。车站主体受北侧220kV高压线、电力管廊和南侧安置房控制，站台宽度11m，有效站台长度120m，岛式地下2层站，线间距14m，车站外边距离高压线7.44m。建筑面积约16923.41m²。车站外包总长308.3m，宽度20.3m。翠园路规划标高+8.3~8.7m，东沙站中心里程轨面标高-8.950m，车站埋深约19.82m。本方案设置4个出入口，分别位于翠园路的南北两侧。

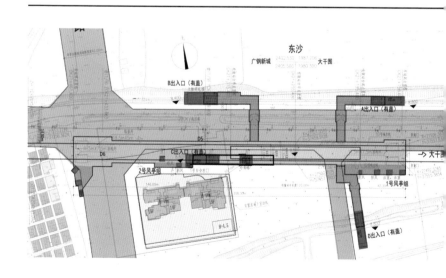

总平面图

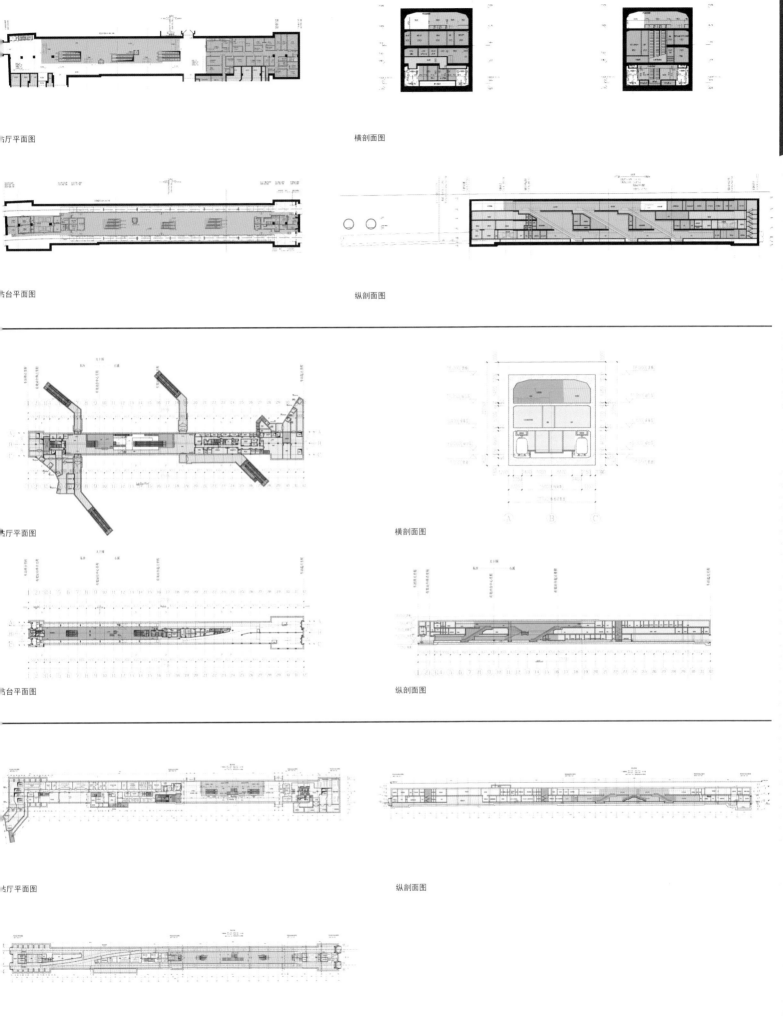

站厅平面图　　　　　　　　　　横剖面图

站台平面图　　　　　　　　　　纵剖面图

站厅平面图　　　　　　　　　　横剖面图

站台平面图　　　　　　　　　　纵剖面图

站厅平面图　　　　　　　　　　纵剖面图

站台平面图

广州市轨道交通十号线
广钢新城站
Guangzhou Metro Line 10
Guanggang New Town Station

广钢新城站东联东沙站，西联西朗站，引导产业片区发展，与规划的佛山十一号线站换乘，加强广佛轨道交通衔接，推动广佛同城化发展进程。广钢新城站为地下2层双岛四线车站，与佛山十一号线"L字"换乘，站前、站后设交叉渡线。车站共设置4个出入口、4组风亭、1个冷却塔。设计总长度395.3m，站台宽17m，标准段宽45m，有效站台长120m。车站总建筑面积40046.75m²，主体建筑面积35996.92 m²，附属建筑面积4049.83m²。

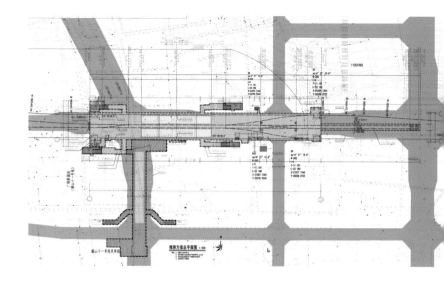

总平面图

广州市轨道交通十号线
西朗站
Guangzhou Metro Line 10
Xilang Station

站位选址在花地大道南北侧，位于一号线西朗既有站进站平台位置。车站东临西朗公交总站，西联一号线西朗站既有车站建筑。西朗站为地上2层、地下2层侧式车站，与广佛线西朗站、一号线、二十二号线西朗站通道换乘。站前、站后分别设单渡线及交叉渡线。车站共设置2个出入口、4组风亭、1个冷却塔。设计总长度473.45m，站台宽5m，标准段宽29.45m，有效站台长120m。车站总建筑面积 27191.82m²，主体建筑面积25117.45m²，附属建筑面积2074.34m²。

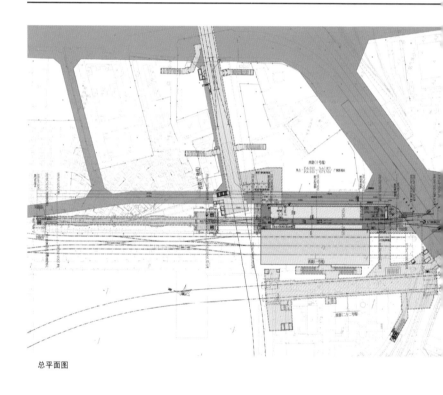

总平面图

广州市轨道交通八号线北延段
东晓南站、同福西站
Guangzhou Metro Line 8 North Extension
Design of Dongxiaonan Station
& Tongfuxi Station

东晓南站位于东晓南路与侨港路交叉口处，车站跨南洲北路，沿东晓南路走向布置。为避开地下大口径管线，减少管线迁改，将原站位往东侧移，有利于改善线路转弯条件，避开两端立交高架桥墩。负一层为站厅层，负二层为站台层，出入口及风亭结合布置于周边地块，与相邻建筑地下室合建。

同福西站位于洪德路与同福西路交叉口南处，车站沿洪德路走向布置，为减少地下管线迁改费、房屋拆迁费，将原站位往东北侧移，降低有效站台轨面标高，调整为南、北端线间距，两站台线路成喇叭口状，有利于改善北端区间线路转弯，避开两端立交高架桥墩，有利于北端区间隧道盾构施工安全过珠江。地下建筑为地铁配套用房，负一层为站厅层，负二层为过厅层，负四层为站台层，出入口及风亭结合布置于周边地块。

总平面图

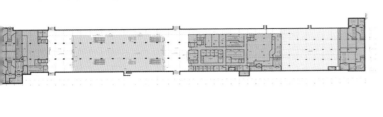
站厅平面图

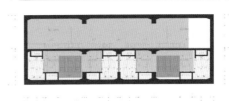
横剖面图

站台平面图

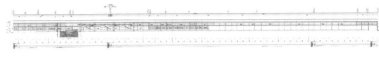
纵剖面图

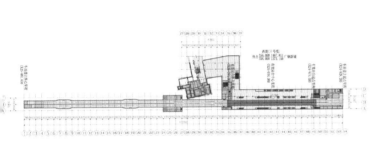
站厅平面图

横剖面图

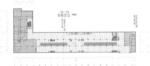
站台平面图

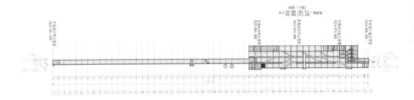
纵剖面图

横剖面图

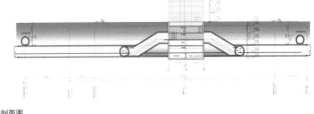
纵剖面图

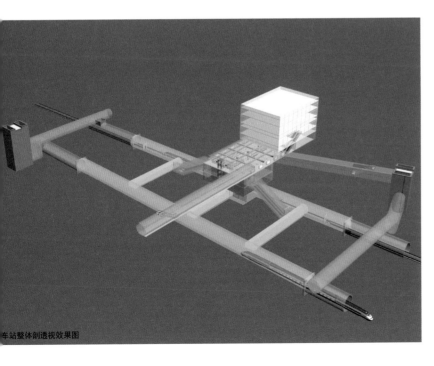
车站整体剖透视效果图

106–117	广州市轨道交通五号线 坦尾站	
	Guangzhou MetroLine 5 Tanwei Station	
118–123	广州市轨道交通二十一号线 山田站、朱村站	
	Guangzhou Metro Line 21 Shantian and Zhucun Stations	
124–125	广州市轨道交通二十一号线 金坑站（绿色建筑三星）	
	Guangzhou Metro Line 21 Jinkeng Station (Green Building 3 Star)	
126–129	广州市轨道交通二十一号线 长平站	
	Guangzhou Metro Line 21 Changping Station	
130–131	广州市轨道交通二十一号线 镇龙站外立面方案设计	
	Guangzhou Metro Line 21 Building Facade Design and Public Area Decoration Design of Zhenlong Station	
132–135	广州市轨道交通四号线 石碁站	
	Guangzhou Metro Line 4 Shiqi Station	
136–139	广州市轨道交通四号线 海傍站	
	Guangzhou Metro Line 4 Haibang Station	
140–143	广州市轨道交通四号线 东涌站	
	Guangzhou Metro Line 4 Dongchong Station	
144–147	广州市轨道交通四号线 低涌站	
	Guangzhou Metro Line 4 Dichong Station	
148–151	广州市轨道交通四号线 黄阁汽车城站	
	Guangzhou Metro Line 4 Huangge Autocity Station	
152–155	广州市轨道交通五号线 西场站建筑外立面设计	
	Guangzhou Metro Line 5 Building Facade Design of Xichang Station	
156–159	广州市轨道交通五号线 西村站建筑外立面设计	
	Guangzhou Metro Line 5 Building Facade Design of Xicun Station	
160–161	广州市轨道交通五号线 区庄站建筑外立面设计	
	Guangzhou Metro Line 5 Building Facade Design of Ouzhuang Station	
162–173	无锡地铁一号线 高架段车站建筑外立面方案设计	
	Wuxi Metro Line 1 Building Facade Design of Elevated Section Station	
174–175	广州市轨道交通广佛线 蠡岗站地面车站建筑方案设计	
	Guangzhou Metro Guangfo Line Ground Station Architectural Design of Leigang Station	

第二部分 高架及地面车站
Elevated and Ground Station

176–177	广州市轨道交通四号线南延段 高架车站建筑方案设计（投标方案）	
	Guangzhou Metro Line 4 Architectural Design of South Extension Elevated Station (Bidding Proposal)	
178–179	广州市轨道交通十三号线 象颈岭站、夏园站车站建筑外立面设计方案	
	Guangzhou Metro Line 13 Building Facade Design of Xiangjingling & Xiayuan Stations	
180–183	佛山地铁二号线 高架段车站建筑外立面方案设计	
	Foshan Metro Line 2 Building Facade Design of Elevated Section Station	
184–185	无锡地铁二号线 高架段车站建筑方案设计（投标方案）	
	Wuxi Metro Line 2 Architectural Design of Elevated Section Station (Bidding Proposal)	
186–187	宁波地铁二号线 高架段车站建筑方案设计（投标方案）	
	Ningbo Metro Line 2 Architectural Design of Elevated Section Station (Bidding Proposal)	
188–189	成都地铁 东洪站、经平院站建筑方案设计（投标方案）	
	Chengdu Metro Architectural Design of Donghong & Jingping yuan Stations (Bidding Proposal)	
190–191	澳门轻轨一期 车站方案设计	
	Macao Light Rail Phase 1 Bidding Proposal Design of Stations	
190–191	广州市轨道交通六号线二期 高架车站建筑方案设计（投标方案）	
	Guangzhou Metro Line 6 Phase 2 Architectural Design of Elevated Station (Bidding Proposal)	
190–191	广州市轨道交通四号线 官桥站、庆盛站建筑方案设计（投标方案）	
	Guangzhou Metro Line 4 Architectural Design of Guanqiao & Qingsheng Stations (Bidding Proposal)	
192–193	北京地铁亦庄线 高架段车站建筑方案设计（投标方案）	
	Beijing Metro Yizhuang Line Architectural Proposal Design of Elevated Station (Bidding Proposal)	
192–193	北京地铁六号线 高架段车站建筑方案设计（投标方案）	
	Beijing Metro Line 6 Architectural Proposal Design of Elevated Station (Bidding Proposal)	
192–193	东莞地铁一号线 高架站外立面方案设计（投标方案）	
	Dongguan Metro Line 1 Building Facade Design of Elevated Section Stations (Bidding Proposal)	
194–195	广州市轨道交通九号线 地面车站建筑方案设计	
	Guangzhou Metro Line 9 Architectural Proposal Design of Ground Stations	
194–195	广州市轨道交通二十一号线 高架段车站方案设计（中标方案）	
	Guangzhou Metro Line 21 Bidding Proposal Design of Elevated Section Station (Bid–Winning Proposal)	

高架车站实景

广州市轨道交通五号线
坦尾站
Guangzhou Metro Line 5
Tanwei Station

车站位于广州市荔湾区大坦沙岛,是全国首个高架站与地下站换乘车站。其中五号线车站为高架车站,车站采用5B编组。

车站总建筑面积约14200m²,其中高架部分建筑面积约7300m²。采用8m岛式站台,有效站台长度106m。六号线为地下一层侧式站台,采用4L编组。

车站周边的道路空间复杂,用地条件极为紧张,高架道路与市政道路均紧邻车站用地。为充分利用场地周边条件,车站将非付费区设置于车站外的入口处,利用大悬挑雨棚提供站外购票空间。并将高架道路下方的空间作为车站的付费区及换乘厅,同时结合轨道与高架道路的空隙形成付费区内的采光天窗,为高架桥下付费区空间提升空间品质。

站台采用开敞式空间设计。金属屋面利用站台电梯筒作为支撑结构,形成下无柱站台空间。金属屋面通过顶面标高变化,形成弧形屋面天际线,结合属屋面立体桁架,使车站金属屋面造型成为车站的标识性元素。

高架及地面车站

高架车站实景

车站局部实景

高架车站实景

高架车站局部实景

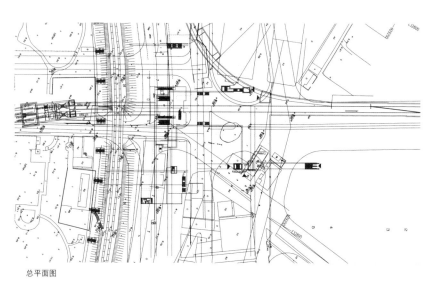

总平面图

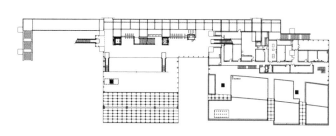

夹层站厅及设备房平面图

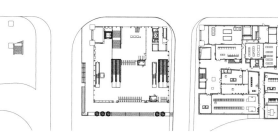

地面站厅及设备房平面图

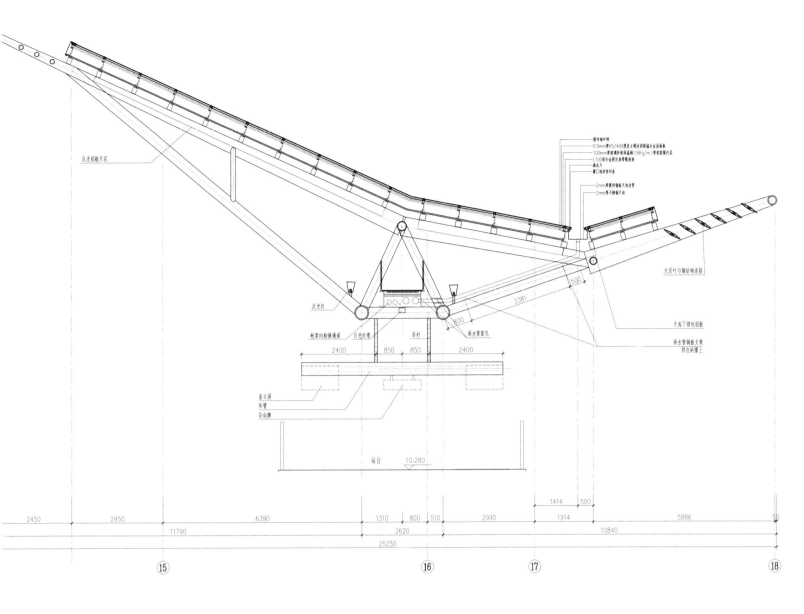

面桁架大样图

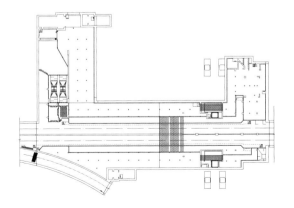

下站台板下层平面图

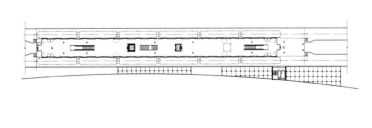

站台平面图

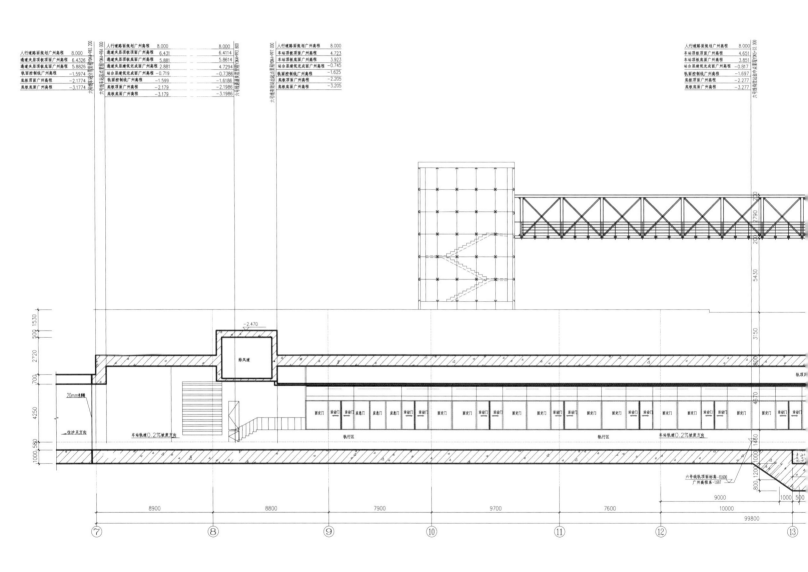

横剖面图

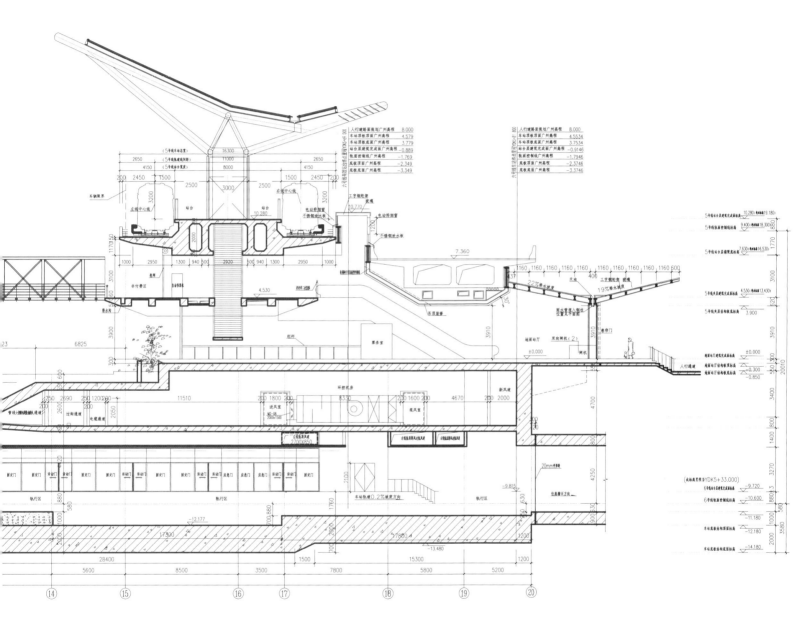

低点实景

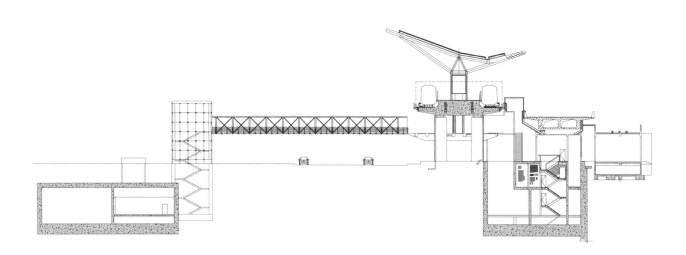

横剖面图

实景

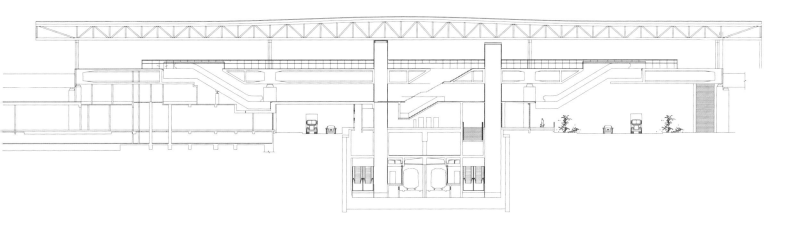
剖面图

车站夜景实景

人行天桥实景

屋面桁架局部实景

轨道及屋面桁架实景

屋面桁架局部实景

车站换乘入口实景

地铁站站厅入口实景

高架站台实景

车站低点实景

广州市轨道交通二十一号线
山田站、朱村站
Guangzhou Metro Line 21
Shantian and Zhucun Stations

车站于2018年12月开通。山田站、朱村站均位于广州市增城区，为二十一号线路中高架车站，车站首层为架空道路，二层为站厅、三层为站台，通过四座天桥与周边人行道联系。车站站台为双岛四线，采用快慢线的行车模式。

车站以"简"作为设计概念，通过造型传达出科技、现代的车站形象。通过横线条的运用，结合立面、顶棚的泛光处理，使得车站能够在白天及夜晚都具备强烈的交通建筑应有的标识特征。

车站站台充分敞开，并利用天光，在轨行区上方分别设置了矩形和菱形窗，并利用格栅为天窗滤光，使天然光柔和地进入站台，为站台提供了舒适光环境。

低点实景

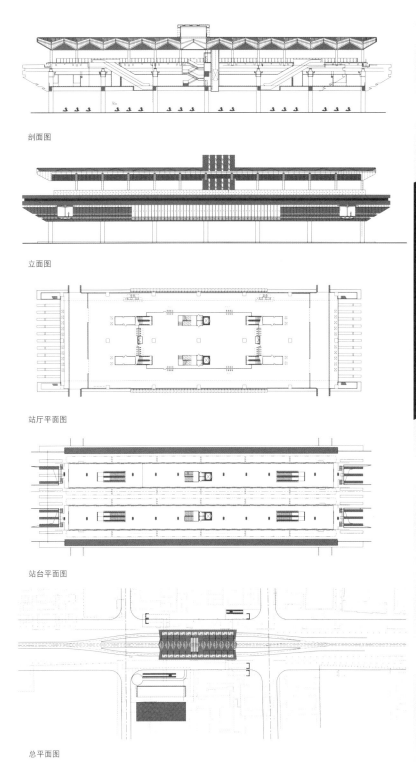

剖面图

立面图

站厅平面图

站台平面图

总平面图

高架及地面车站

低点实景

车站局部实景

车站鸟瞰实景

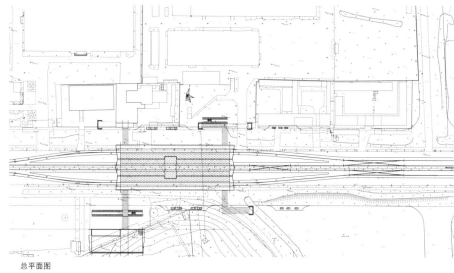

总平面图

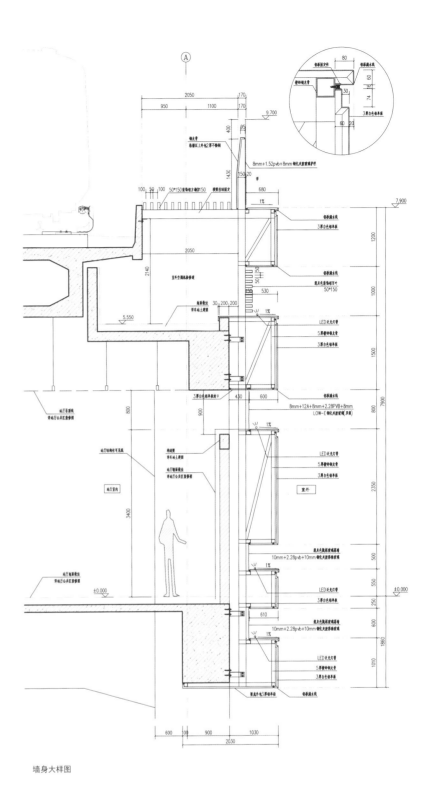

墙身大样图

车站鸟瞰实景

车站局部实景

车站鸟瞰实景

车站站台实景

车站站厅实景

车站低点实景

鸟瞰实景

广州市轨道交通二十一号线
金坑站（绿色建筑三星）
Guangzhou Metro Line 21
Jinkeng Station (Green Building 3 Star)

金坑站位于广州市广汕公路以南的黄屋村与银岭村交界处，线路沿广汕公路南边地块敷设，车站呈东西向布置。车站为地面高架两层10.5m岛式站台，全长170.4m，标准段宽38.8m，车站总高约19.18m。车站总建筑面积约11267.52m^2，其中主体基底建筑面积约6611.52m^2。

车站运用了"三角与起伏"作为基本造型元素，通过材质与高低的变化，塑造出车站顶部的造型变化，既形成独特的空间变化，又通过车站的现代造型元素表达岭南传统文化精髓。材质的运用与高低的变化，塑造了一个极富辨识性的高架站象。车站立面通过幕墙与独特的结构表现，营造出良好的立面效果，车站形统一大气，连贯并富于韵律感。通过立面通透材料的运用，形成灵巧的体量系，降低车站的体量感，减小车站对周边环境的压迫。整个高架站造型简洁充满力量与速度感。

局部实景1

局部实景2

立面图

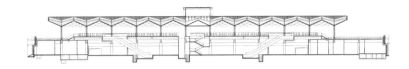

剖面图

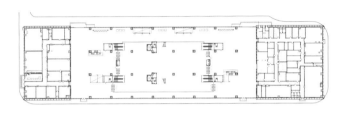

站厅平面图

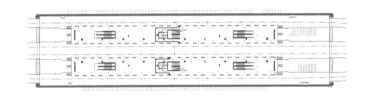

站台平面图

高架及地面车站

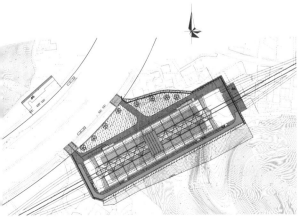

总平面图

鸟瞰实景

鸟瞰实景

广州市轨道交通二十一号线
长平站
Guangzhou Metro Line 21
Changping Station

长平站位于广州市黄埔区永顺大道，地处广州长岭居的门户位置，周边为新开发的商业居住项目。车站总建筑面积约14000m²，采用侧式站台，站厅站台同层布置于车站二层，地面首层为设备用房及预留空间。

车站根据二十一号线高架车站统一的风格，采用简约的横向线条作为立面主要元素，并结合舒展的站厅顶棚，形成标示度极高的车站造型。不同质感的材料天窗柔和的自然光结合，赋予车站空间情感与气质。

车站连接的过街天桥，与车站顶棚连成一体，统一处理，使车站尺度与周边整体环境呼应，塑造车站在长岭居区域的门户形象。

站局部实景

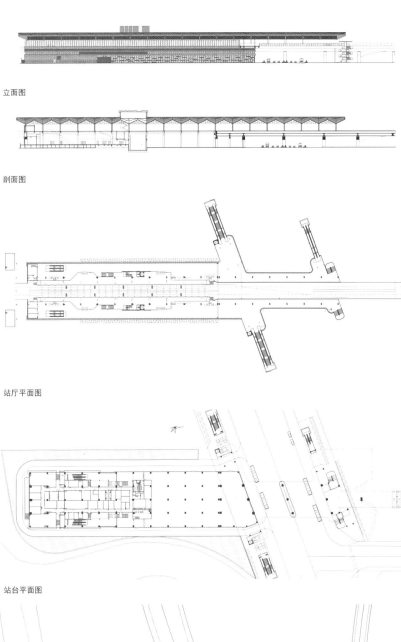

立面图

剖面图

站厅平面图

站台平面图

总平面图

高架及地面车站

站局部实景

车站局部实景2

鸟瞰实景

车站局部实景1

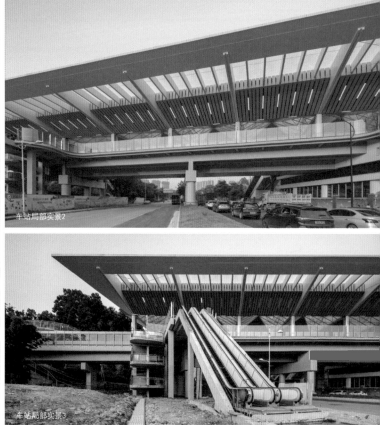
车站局部实景2

车站局部实景3

站台实景1

站台实景2

站台实景3

站台实景4

车站低点实景

广州市轨道交通二十一号线
镇龙站外立面方案设计
Guangzhou Metro Line 21
Building Facade Design and Public Area Decoration Design of Zhenlong Station

车站于2017年12月开通。车站位于广州市黄埔区，为二十一号线与十四号线换乘车站。首层为站厅，二三层为远期功能预留，站台位于地下。

外立面采用简单、干净的横向条作为车站造型的主要元素，并通过横线条的弧线处理，塑造出简洁而具有特征性的立面风格。首层站厅立面以玻璃幕墙为主，为车站空间提供充足阳光的同时，带来立面强烈的虚实对比。顶部造型呼应元素，打造出弧形的车站天际线，使整个车站与周边的山脉环境取得对话。

站厅实景

站厅实景

局部实景

站台实景

高架及地面车站

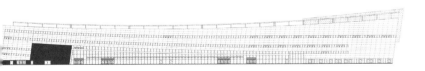
立面图3

立面图4

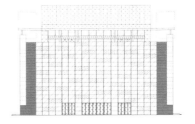
立面图2

立面图1

车站低点实景

广州市轨道交通四号线
石碁站
Guangzhou Metro Line 4
Shiqi Station

石碁站于2006年开通。车站位于广州市番禺区,是广州第一条轨道交通高架线路。站址位于60m宽市莲路与京珠高速公路立交口东南侧,站位东侧为广东女子职业技术学院,西侧为京珠高速公路。总建筑面积约6600m²。线间距4.0m,车站设侧式4m站台。车站长度75m,总宽度45.60m,有效站台长度72m。站台层轨面高度15.900m(广州高程),车站采用4L编组。

站厅设置于地面层,设南北两个出入口。二层为两个侧式站台。设备管理用外置于站台外侧,车站底层部分架空,为公交系统客流留出接驳空间。车站采用圆弧形造型,造型在四号线高架站统一中寻求变化,形成线路车站确的标识性。

高架及地面车站

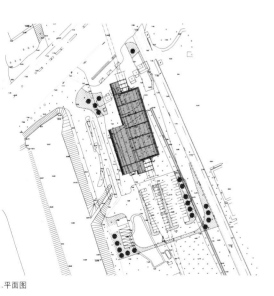
平面图

鸟瞰实景

车站低点实景

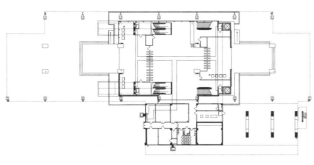

站厅平面图

站台平面图

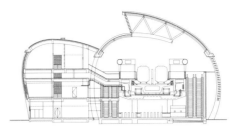

剖面图1

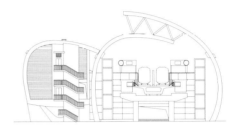

立面图

车站室内局部实景

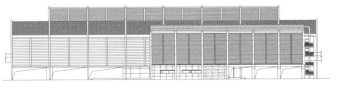

立面图

半鸟瞰航拍实景

广州市轨道交通四号线
海傍站
Guangzhou Metro Line 4
Haibang Station

海傍站于2006年开通，是广州第一条轨道交通高架线路。车站位于广州市番禺区长南路，周边为广州亚运城媒体中心。车站采用4L编组，总建筑面积约5300m²。车站长度75m，总宽度45.60m，有效站台长度72m，站台层轨面高度14.400m（广州高程）。

站厅设置于地面层，设南、北、东三个出入口。二层为两个侧式站台。设备理用房外置于站台外侧。车站采用圆弧形造型，首层站厅通过大面积悬挑棚，形成架空式入口广场。广场雨棚下的架空空间，为车站的人流集散提供良好的空间条件。

高架及地面车站

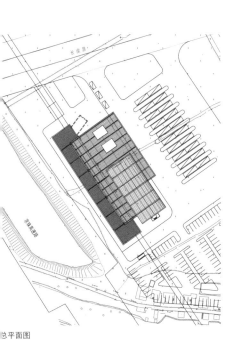

总平面图

鸟瞰实景

鸟瞰实景

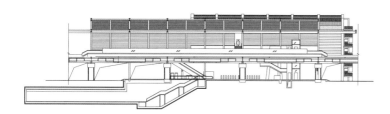
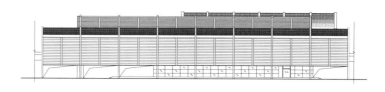

纵剖面图　　　　　　　　　　　　　　　　　　　　立面图

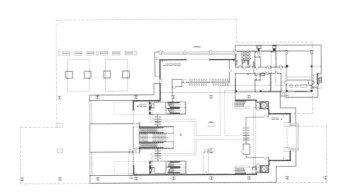

站厅平面图

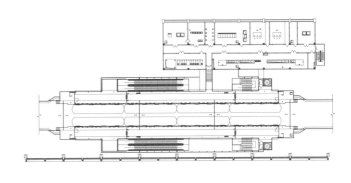

站台平面图

黄立面图

架空式入口广场实景

车站低点实景

广州市轨道交通四号线
东涌站
Guangzhou Metro Line 4
Dongchong Station

东涌站于2006年开通,是广州第一条轨道交通高架线路。车站位于广州市番禺区番禺东涌大道与茂丰路交界处。车站采用4L编组,总建筑面积6509m²,线间距4m,车站采用侧式站台。车站长度77m,宽度23m,有效站台长度72m,站台层轨面高度13.900m(广州高程)。

站厅设置于地面层,设南北两个出入口。二层为两个侧式站台。设备管理用房外置于站台外侧,车站底层部分架空,为公交系统客流留出接驳空间。车站在号线高架车站统一的弧形造型语言下,结合车站入口处弧形的架空入口空间形成车站入口的标识性,同时在统一的车站形式中,表达出车站自有的个性征,在统一中寻求变化。

高架及地面车站

车站鸟瞰实景

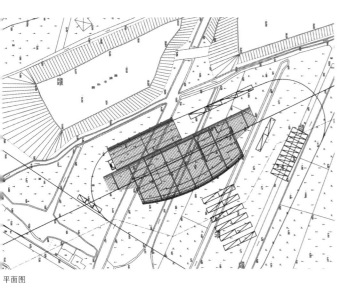
平面图

车站鸟瞰实景

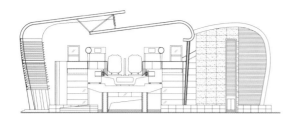
横剖立面图

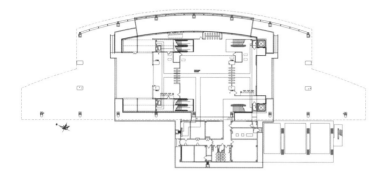
站厅平面图

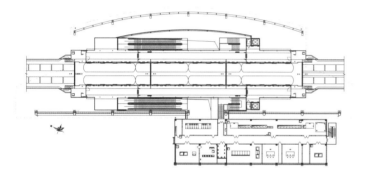
站台平面图

车站低点实景

车站局部实景

车站站厅室内实景

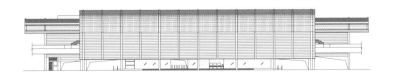

纵立面图

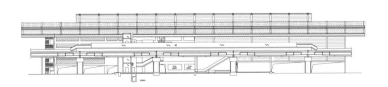

纵剖面图

车站低点实景

广州市轨道交通四号线
低涌站
Guangzhou Metro Line 4
Dichong Station

低涌站于2006年开通，位于广州市番禺区，是广州第一条轨道交通高架线路。站址位于60m宽规划南德路与50m宽规划前清路路口西南端的地块中，京珠高速公路立交口东南侧，站位西侧为京珠高速公路。

车站采用4L编组，总建筑面积约6300m²。线间距4.0m，车站设侧式4m站台。车站长度100.4m，总宽度43.10m，有效站台长度72m。站台层轨面高度13.600m（广州高程），站厅设置于地面层，设南北两个出入口。二层为两个侧式站台，备管理用房外置于站台外侧，车站底层部分架空，为公交系统客流留出接驳空间。车站采用圆弧形造型，造型在四号线高架站统一中寻求变化，形成线路站明确的标识性。

车站鸟瞰实景

高架及地面车站

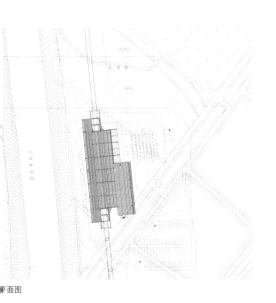
平面图

车站鸟瞰实景

145

车站低点实景

站厅室内实景

站厅局部实景

站厅局部实景

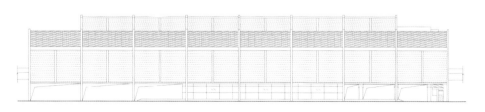

立面图1

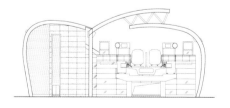

立面图2

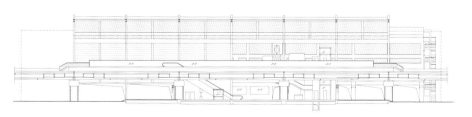

纵剖面图

高架车站低点实景

广州市轨道交通四号线
黄阁汽车城站
Guangzhou Metro Line 4
Huangge Autocity Station

黄阁北站于2006年开通，是广州第一条轨道交通高架线路。车站位于广州市南沙区黄阁北路与市南路交界处。车站采用4L编组，总建筑面积6100m²。线间距4.0m，车站设侧式4m站台。车站长度77.7m，有效站台长度72m。站台层轨面高度19.940m（广州高程）。

该站为路中高架站，车站体量整体位于道路上方。为减少对道路的压迫感，站厅设置于二层，结合出入口及设备用房，分设于道路的两侧，减少车站在道路的体量。道路两侧站厅通过天桥与路中站台进行联系。车站主体底层为架面，二层为天桥，三层为两个侧式站台，位于道路正上方。

车站造型利用弧形的车站元素，让道路上方的车站取得飘逸、灵动、轻巧型效果，避免了对道路产生不利的景观效果。

高架及地面车站

车站鸟瞰实景

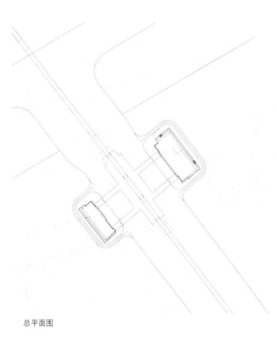
总平面图

车站鸟瞰实景

车站鸟瞰实景

低点局部实景

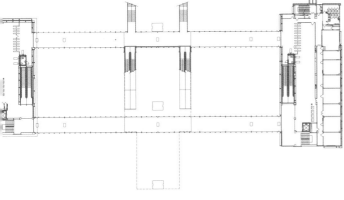
厅平面图

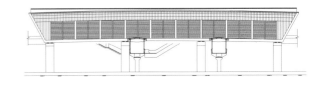
立面图

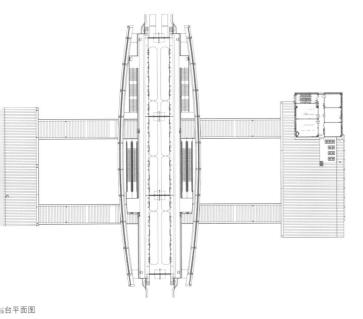
站台平面图

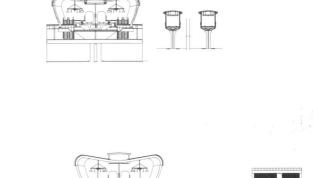
横剖面图

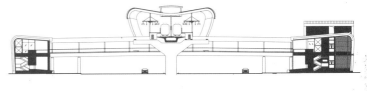
横剖面图

站台空间实景

车站换乘空间实景

车站鸟瞰效果图

广州市轨道交通五号线
西场站建筑外立面设计
Guangzhou Metro Line 5
Building Facade Design of Xichang Station

西场站车站站厅设置于地面层，分散布置于地面的周边地块。建筑立面设计结合周边环境，将建筑体量化整为零，通过体块的穿插变化，形成站厅建筑形态的丰富变化，同时，使车站更好地融入周边环境中。外立面明确的色彩，为车站带来强烈的可识别性，乘客能够便捷地在周边复杂的环境中找到车站。受站点设置与站点周边环境条件的制约，车站站厅设于地面，组合了其他城市公用设施（如公共卫生间、变电所、居委、商铺、设备房等），形成功能高度浓缩的小型城市交通综合体，建筑主体随其周边城市环境呈现自我个性的建筑表达，注重现代简洁的量组合与穿插，并通过材料搭配对建筑细节精心刻划，打造属于新时代的交建筑品位。

高架及地面车站

车站低点实景

车站鸟瞰实景

立面图1

立面图2

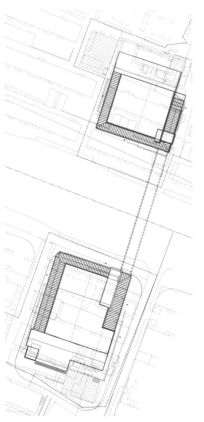
总平面图

车站低点实景

车站局部实景

车站局部实景

车站半鸟瞰效果图

广州市轨道交通五号线
西村站建筑外立面设计
Guangzhou Metro Line 5
Building Facade Design of Xicun Station

1. 分隔而不分散
由于线路条件的限制，需在东风路两侧设置分隔的站厅，而车站使用功能则需要强化整体的建筑形象，因此采用两侧一致的设计手法，而两边站厅的具体功能差别（南站厅为主设备用房区），使建筑造型在相似中又呈现差异性。

2. 低层而不低调
站厅周边皆为9层左右的住宅，站厅主体高度比周边建筑要低，而其本身公共建筑特性要求其造型须具备一定标志性，为乘客提供明确的使用导向，因此设计采用了体量分明的体块叠加穿插手法形成具有视觉冲击力的造型效果。

3. 简约而不简单
造型设计以广泛使用的石材、玻璃及金属铝板的材料搭配，色彩设计则以（百叶框架）、白（玻璃）、灰（石材）、黄（铝板）为主调，有效打破外空间的分离感，加以精心的细节设计，使简单的建筑功能创造出简约的建筑效果。

总平面图

高架及地面车站

车站低点实景

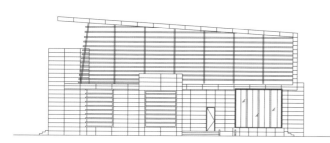
立面图1

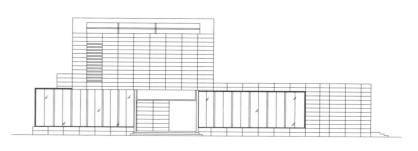
立面图2

车站局部实景

车站局部实景

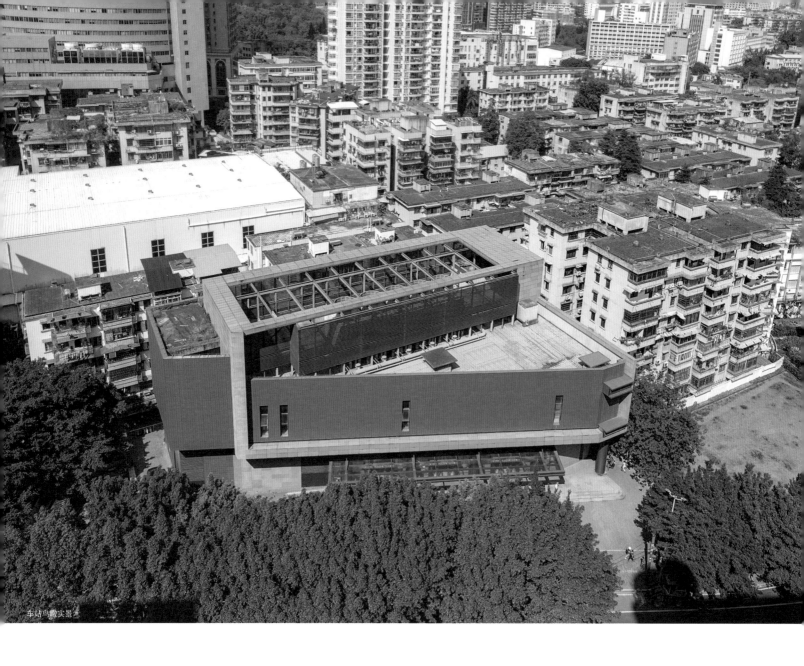

车站鸟瞰实景

广州市轨道交通五号线
区庄站建筑外立面设计
Guangzhou Metro Line 5
Building Facade Design of Ouzhuang Station

区庄站位于广州市环市路，站厅层设置于地面首层，地下布置站台层。建筑立面根据周边紧张的环境，对城市道路及周边建筑作出呼应。体型上通过体量的划分，减少建筑尺度，与周边高楼环境建立良好关系，让建筑融入周边的建筑环境中。

建筑立面材料考虑节能，采用环保性较好的陶土板作为主要的立面材料，减环境污染。简洁的立面与局部的建筑细部形成良好的尺度关系，丰富了建筑情，同时也形成了交通建筑功能性强、简约、现代的风格特征。

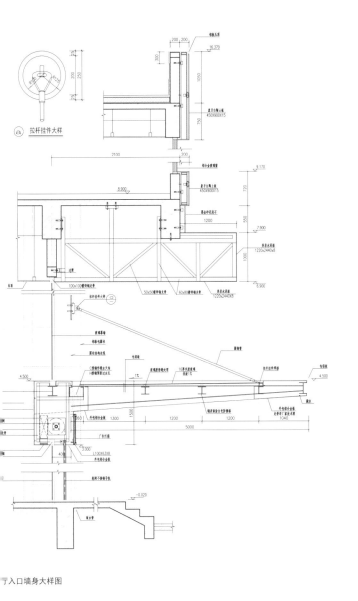
丁入口墙身大样图

车站低点实景

高架及地面车站

口局部实景

车站低点实景

车站低点实景

无锡地铁一号线
高架段车站建筑外立面方案设计
Wuxi Metro Line 1
Building Facade Design of Elevated Section Station

无锡作为著名的江南文化名城,在一号线高架站的造型设计中,吸收江南文化的精髓,把江南文化与地铁交通功能紧密结合,设计中提出"一式异景"的设计原则。车站是用一种地域风格和构成模式,通过不同的文化背景打造一条线路中的不同车站。车站充分考虑当地的气候特征及本土产业——无锡光伏玻璃板的利用,形成独特的"无锡模式"地铁文化标杆。

车站通过分析周边环境的特性,在统一的造型语言下,通过局部特殊的处理形成对环境的呼应,并形成每个车站自身的特点,让全线高架车站形成完整统一的形象风格。

高架及地面车站

面图

站低点实景

车站低点实景

车站局部实景

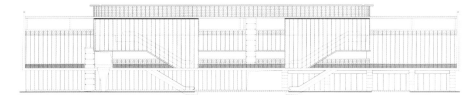

立面图

车站低点实景

车站低点实景

车站实景

车站实景

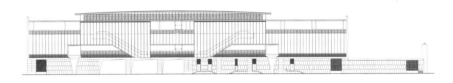

立面图

车站实景

车站低点实景

车站实景

车站实景

招标阶段

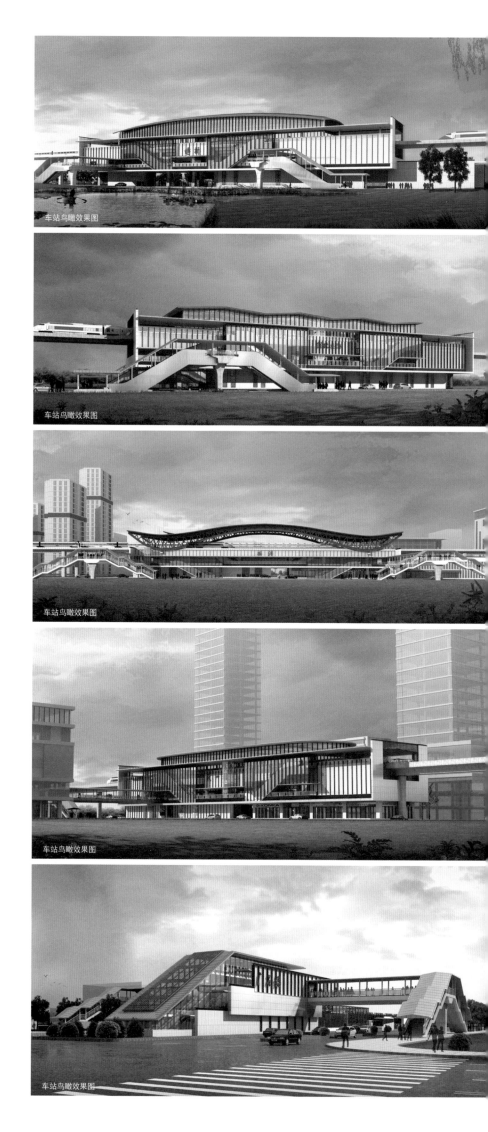

车站鸟瞰效果图

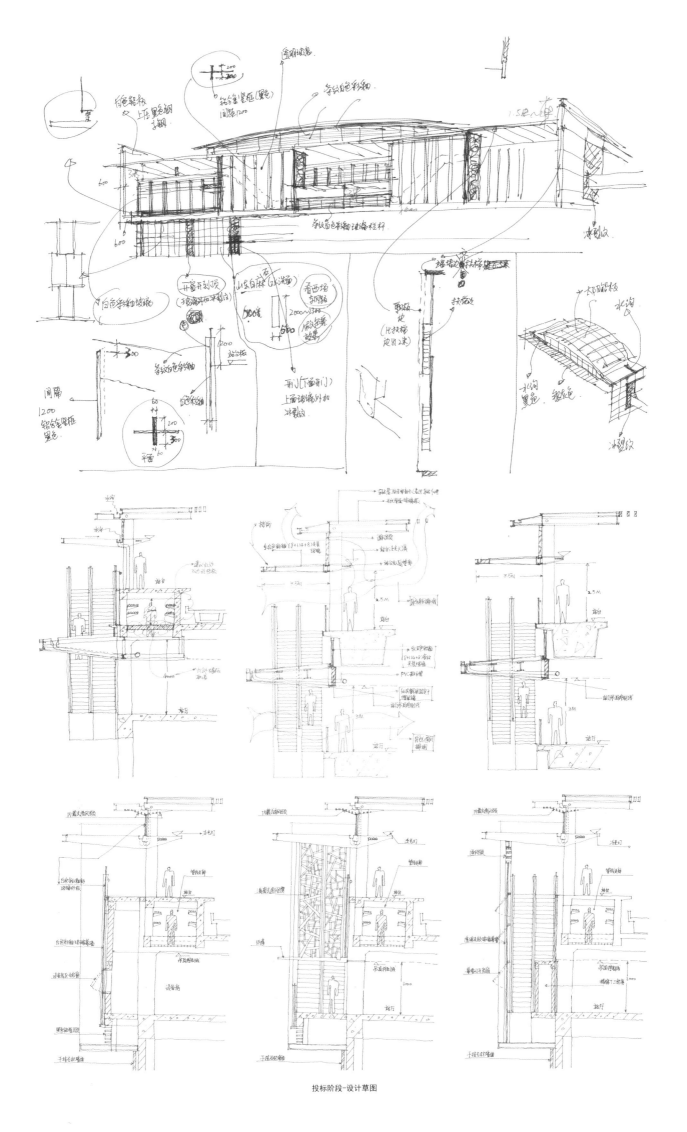

投标阶段-设计草图

投标阶

车站鸟瞰效果图

车站鸟瞰效果图

车站鸟瞰效果图

车站鸟瞰效果图

车站鸟瞰效果图

汽车文化
（临近多间汽车4S店）

居住文化
（临近高档居住区）

水乡文化
（临近锡北运河）

村庄文化
（临近村落）

商业文化
（临近繁荣商业）

中标）

堰桥站

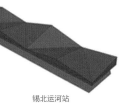
锡北运河站

西漳站

天一路站

广石路

车站鸟瞰效果图

车站鸟瞰效果图

车站鸟瞰效果图

车站鸟瞰效果图

车站鸟瞰效果图

车站鸟瞰效果图

广州市轨道交通广佛线
蠕岗站地面车站建筑方案设计
Guangzhou Metro Guangfo Line
Ground Station Architectural Design of Leigang Station

广佛线蠕岗站为地下两层岛式车站，站后设单渡线。车站位于规划中的南海区蠕岗公园东侧车站，采用明挖两层框架结构，中柱岛式站台，地面出入口车站设备房设置于地面，结合广佛线主变电所设置，同时把站厅设置于地面，减少地下建设面积，缩减车站规模。

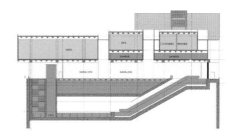

车站剖面图

车站站厅平面图

站台效果图

高架及地面车站

厅室内效果图

车站低点效果图

鸟瞰效果图

广州市轨道交通四号线
南延段高架车站建筑方案设计（投标方案）
Guangzhou Metro Line 4
Architectural Design of South Extension Elevated Station (Bidding Proposal)

高架站造型设计上结合车站交通空间及功能特点，以鲜明的白色线条作为建筑的设计母题，通过造型元素暗示车站交通客流组织方向，较好地解决高架站台自然通风采光和遮风挡雨的问题，用理性严谨的方式打造一个既简捷、现代又功能化的交通建筑。

低点效果图

低点效果图

鸟瞰效果图

高架及地面车站

室内效果图

站台空间效果图

鸟瞰效果图

低点效果图

广州市轨道交通十三号线
象颈岭站、夏园站车站建筑外立面设计方案
Guangzhou Metro Line 13
Building Facade Design of Xiangjingling & Xiayuan Stations

广州地铁十三号线地面站在建筑选材与设计风格上追求工业化、模数化，通过建筑"简·捷"的设计手法与造型形态表达现代交通建筑的交通属性，既注重交通建筑以功能为主导的设计原则，又通过整体风格的把握与个性细部的推敲，研究出具有地域历史特征的人文元素，使车站从功能上升为地铁文化的层面。

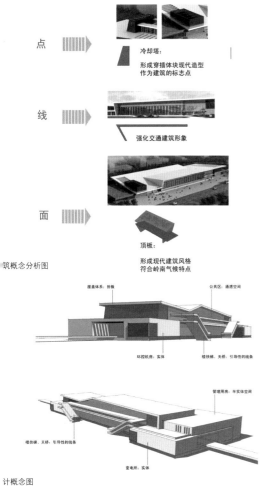

鸟瞰效果图

佛山地铁二号线
高架段车站建筑外立面方案设计
Foshan Metro Line 2
Building Facade Design of Elevated Section Station

佛山是一个怀着浓厚岭南情怀的城市，无论是历史保留的旧建筑还是现代新建筑，都从不同的角度演绎着各种岭南气息，而佛山二号线作为市区第一条设高架站的地铁线，其车站对城市的意义成了高架站造型设计的切入点。结合佛山岭南特色，以"龙舟"作为车站设计的主题，打造犹如"龙舟"般的车站，寓意佛山地铁势如破竹的迅猛发展。在技术上，车站通过借鉴岭南民居的地域特性，把其原理在车站的通风遮阳挡雨策略中，使车站无论从外形还是内涵，都是一个深含岭南精髓的现代化交通建筑。

高架及地面车站

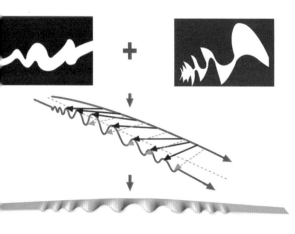
筑屋盖概念设计分析图

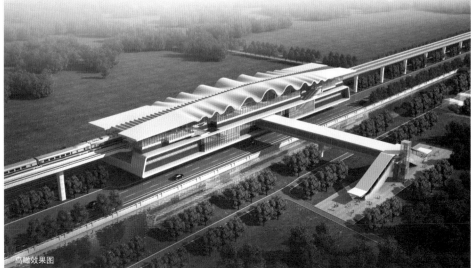
鸟瞰效果图

鸟瞰效果图

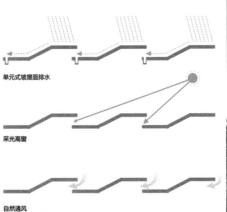

单元式坡屋面排水

采光高窗

自然通风

低点效果图

岛式站低点效果图

无锡地铁二号线
高架段车站建筑方案设计（投标方案）
Wuxi Metro Line 2
Architectural Design of Elevated Section Station (Bidding Proposal)

无锡二号线高架站造型设计上延续着一号线"一式异景"的原则，以车站交通功能为核心，进一步强化线路江南文化的大共性，提炼出不同站点的地域文化并表达于车站的标志性造型中，在车站细节上同样关注地域性气候的应对与本土产业光伏玻璃的使用，致力进一步打造无锡地铁专属的高品位文化品牌。

侧式车站低点效果图　侧式车站鸟瞰效果图
侧式车站低点效果图　侧式车站鸟瞰效果图
侧式车站低点效果图　侧式车站鸟瞰效果图
侧式车站低点效果图　侧式车站鸟瞰效果图
侧式车站低点效果图　侧式车站鸟瞰效果图

高架及地面车站

岛式车站鸟瞰效果图

车站低点效果图

车站鸟瞰效果图

站台空间效果图

宁波地铁二号线
高架段车站建筑方案设计（投标方案）
Ningbo Metro Line 2
Architectural Design of Elevated Section Station (Bidding Proposal)

本项目以"C+T"模式研究高架地铁的功能与造型，C代表文化，T代表技术。高架地铁站以体现江南水乡文化与地段的地域文化作为设计元素，把四个路段的站点归纳为商业属性、文化属性、交通属性与地域属性，然后创造出高架站不同的顶部造型，体现个性的同时突出共性。

技术与功能安排上完全符合现代的高架地铁站的需求，体现现代交通建筑的洁与工艺，同时对人流的方向做出充分研究，契合现代交通建筑以人为本的计理念。

T(Technology) + C(Culture)

路林市场站
商业属性

双桥站
交通属性

宁波大学站
文化属性

东外环路站
地域属性

盖概念设计

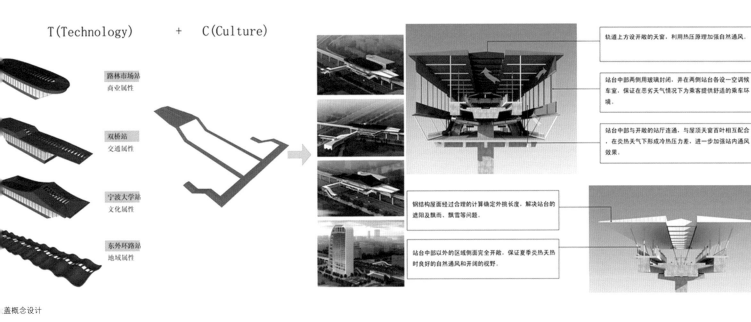

高架及地面车站

车站低点效果图

车站低点效果图

车站低点效果图

车站概念设计效果图

站厅概念设计效果图　　站台概念设计效果图

成都地铁
东洪站、经平院站建筑方案设计（投标方案）
Chengdu Metro
Architectural Design of Donghong & Jingping yuan Stations (Bidding Proposal)

成都地铁经平院站结合规划路，车站与设备房围合成一个站前广场，以吸引周边的客流，并方便联系规划路的公交站。首层开发商物业，二层为站厅及设备管理用房，三层为站台及牵引变电所。根据成都的气候条件，站厅、站台采用栏杆、雨棚等设计元素。楼扶梯部分采用玻璃体，斜线造型体现现代交通建筑的现代感，屋盖采用弧形轻钢屋面板，设备房叠级布置，立体绿化，各天面通过室外楼梯连接，体现山地建筑特征。

东洪路站采用明挖地下一层侧式站台，地上两层，造型通过简洁的体量关系表达交通建筑的性格，玻璃幕墙与石材墙面形成强烈对比。车站地面首层设一千多平方米的商业开发，并设置地下过街通道。车站东北地块考虑设置社会停车场及公交车站，与车站接驳。

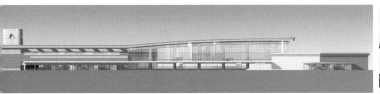
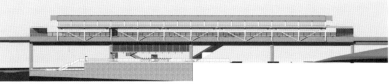

站立面图　　　　　　　　　　　　　　　　岛式车站立面图

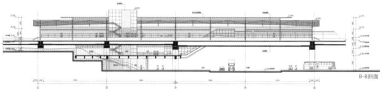

站剖面图　　　　　　　　　　　　　　　　侧式车站剖面图

高架及地面车站

岛式车站效果图

侧式车站效果图

站厅室内效果图　　　　　　　　站台空间效果图

澳门轻轨一期
车站方案设计
Macao Light Rail Phase 1
Bidding Proposal Design of Stations

澳门特别行政区政府于2009年10月落实了轻轨兴建方案，项目包括建立一条主要为高架形式的轻轨路线，少量地下隧道及地下车站；线路从澳门北区关闸开始，经环绕澳门东南区至氹仔，再由西至东绕过路氹新城至澳门国际机场及北安码头，沿线经过澳门的主要关口、住宅区、酒店、娱乐场和商业区。

车站屋盖的构造为钢结构檩条，支撑隐框式铝合金龙骨，面层安装太阳能光伏玻璃，外围2m宽区域为深灰色金属屋面板，站内轨行区上方中心位置开启天窗，天窗与屋面之间安装铝合金防雨百叶。太阳能板用于收集太阳能，转化为电能供应站厅、站台的景观照明，并兼具防风、遮雨、隔热功能。黑色的太阳能板与深灰色的金属屋面板整体安装，美观简洁。光伏屋面有效减少二氧化硫和氮氧化物的排放，实现节能、环保的目标。

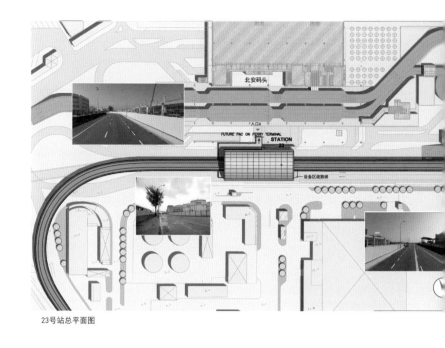

23号站总平面图

广州市轨道交通六号线二期
高架车站建筑方案设计（投标方案）
Guangzhou Metro Line 6 Phase 2
Architectural Design of Elevated Station (Bidding Proposal)

广州市轨道交通六号线的高架车站设计，通过对既有的高架车站设计总结和分析，在设计中提出解决方案。

楼扶梯外装玻璃，将站厅空调空间延伸至站台，彻底解决挡雨问题，为乘客提供舒适的环境，以人为本，端部尽量封闭，尽量减少站台飘雨对乘客及管理人员的影响。造型简洁，较好地运用玻璃与金属材料，视觉通透，空间明快，利用金属飘板遮阳，体现地域特色。站台玻璃中段的金属遮阳板能有效减少屋顶外挑的长度，降低钢结构的造价，同时更好地解决站台的遮阳问题。

鸟瞰效果图

广州市轨道交通四号线
官桥站、庆盛站建筑方案设计（投标方案）
Guangzhou Metro Line 4
Architectural Design of Guanqiao & Qingsheng Stations (Bidding Proposal)

官桥站车站主体在确保施工可行性的前提下，与原有区间变电所有机融合，将其屋面改造为车站站厅，整体造型协调，避免形体过长对城市景观的影响。庆盛站在充分论证分析下，确保施工可行性，移动站位，使车站与火车站呈T形换乘，大大缩短换乘距离，并结合周边公交系统的合理规划设置，以求各系统间无缝接驳，实现轨道交通的最大化。

车站低点效果图

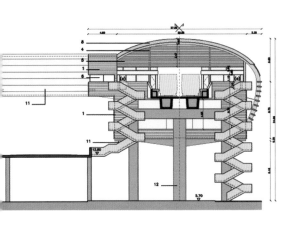

立面图

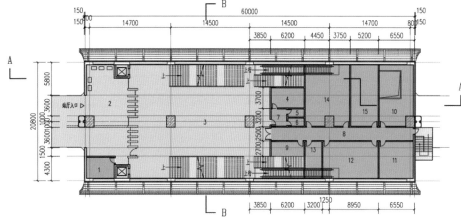

23号站站厅平面图

高架及地面车站

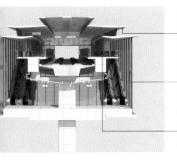

天窗部分飘板的加长，保证了雨天情况下车站内部都能正常通风。

楼扶梯空间外包玻璃在保持视线通透前提下，将站厅（空调空间）进一步延伸至站台，即使在雨天情况下，也能有效保证舒适的乘车环境。

站台玻璃下端设置通风隔栅，与屋顶天窗百叶相互配合，在炎热天气下形成冷热压力差，进一步加强站内通风效果。

璃中段的金属遮阳板能有效地减少屋顶长度，降低钢结构的造价，同时更好地合的遮阳问题。

璃外窗采用电动控制开启，平时打开通长时电动关闭。

栅与天窗形成热压作用的空气流动，加合的通风效果，即使在雨天也不会受到

解图

低点效果图

鸟瞰效果图

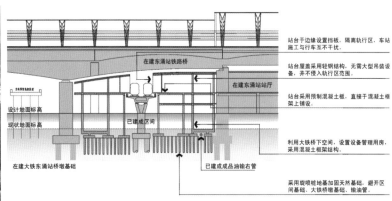

站台于边缘设置挡板，隔离轨行区，车站施工与行车互不干扰。

站台屋盖采用轻钢结构，无需大型吊装设备，并不侵入轨行区范围。

站台采用预制混凝土板，直接于混凝土框架上铺设。

利用大铁桥下空间，设置设备管理用房，采用混凝土框架结构。

采用旋喷桩地基加固天然基础，避开区间基础、大铁桥墩基础、输油管。

推荐方案可行性研究

北京地铁亦庄线
高架段车站建筑方案设计（投标方案）
Beijing Metro Yizhuang Line
Architectural Proposal Design of Elevated Station (Bidding Proposal)

该项目探索新的车站功能模式及建筑造型，高度集成设计管理系统，便于使用及管理。建筑形式具有线路标识性，并反映车站个性功能与环境特点。设计从"一线一景"出发，在共性中体现个性。屋顶结合功能作局部变化，并通过各自车站的空间功能的差异，形成整体造型的变化。立面采用中空玻璃幕墙及穿孔铝板，确保车站内部环境的同时，减少光反射对地面车辆的影响。

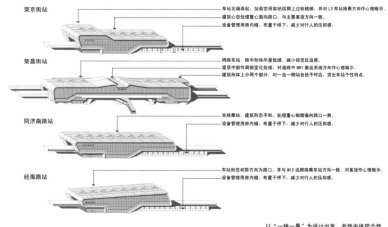

以"一线一景"为设计出发，共性中体现个性

车站形体分析图

北京地铁六号线
高架段车站建筑方案设计（投标方案）
Beijing Metro Line 6
Architectural Proposal Design of Elevated Station (Bidding Proposal)

项目强调一线一景的设计目标，提出功能与形式的统一，及共性与个性统一的设计理念。通过节奏与变奏的设计手法，打造车站在统一的形式下不同车站的个性。在屋顶、墙面等局部，通过材料细节的表达，形成车站颜色与材质的局部个性，为线路整体形象提供丰富的变化。

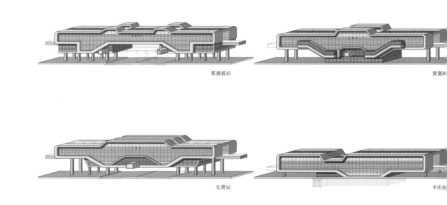

车站形体分析图（侧式）

东莞地铁一号线
高架站外立面方案设计（投标方案）
Dongguan Metro Line 1
Building Facade Design of Elevated Section Stations (Bidding Proposal)

车站方案从波浪的造型中，提取出起伏的屋面造型，并通过屋顶折线强调山体的天际线，站台外遮阳百叶形体强调山体倒影的形体。屋顶剖面的关系也表达出山水概念，形成活跃丰富的站台空间。材质运用主要通过铝板屋面、铝合金百叶、锯齿形玻璃幕墙及混凝土等不同材质对比，形成丰富的立面造型。锯齿形幕墙结合太阳角度，保证站厅的采光，也隔绝了西晒的热量。

道滘东车站鸟瞰效果图

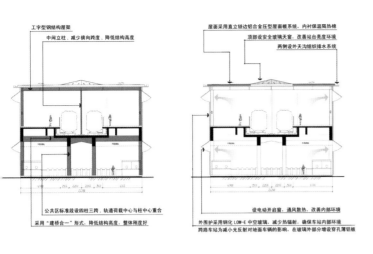
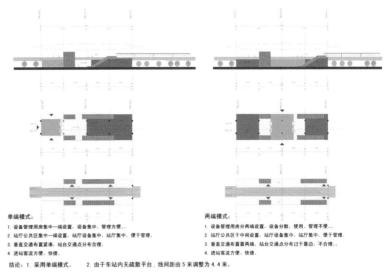

构及围护系统分析图　　　　　　　　　车站功能模式比选确定分析图

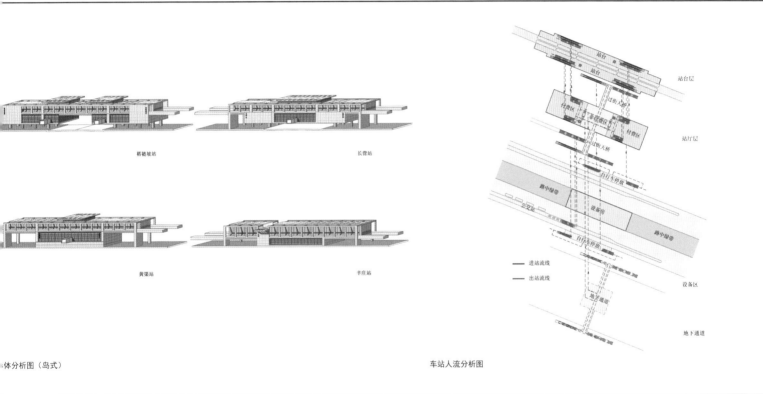

体分析图（岛式）　　　　　　　　　　车站人流分析图

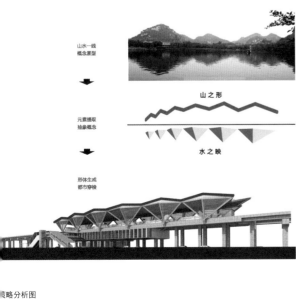

略分析图　　　　　　　　　　　　　　望洪站台空间效果图

高架及地面车站

广州市轨道交通九号线
地面车站建筑方案设计
Guangzhou Metro Line 9
Architectural Proposal Design of Ground Stations

九号线地面车站为高增站,建筑外立面设计采用折板作为主要造型元素,通过建筑形体错动、穿插的手法,强调建筑造型的力量感与动态感,从而表达交通建筑的气质。立面局部采用了红色的色彩元素,强化立面的特质与可识别性。整体建筑体量横向展开,形成舒展、灵动的建筑形态。

方案设计1鸟瞰效果图

广州市轨道交通二十一号线
高架段车站方案设计
(中标方案)
Guangzhou Metro Line 21
Bidding Proposal Design of Elevated Section Station (Bid-Winning Proposal)

二十一号线是联系广州市区及增城区这两个岭南文化代表地区的轨道交通线路。设计从岭南传统建筑屋顶瓦片提取元素,以瓦片铺设的构造形式作为造型的设计主体,表达出"岭"的意象。同时,上下起伏的造型也与线路山地起伏形成呼应。以瓦的造型作为二十一号线车站的出发点,根据站点不同的周边环境,衍生出不同的造型形式,适应场地不同的环境,形成车站与环境相互协调的建筑形式。

双岛方案设计鸟瞰效果图

双岛方案设计低点透视效果图

方案设计2鸟瞰效果图

方案设计2低点效果图

方案设计1低点效果图

高架及地面车站

方案设计鸟瞰效果图

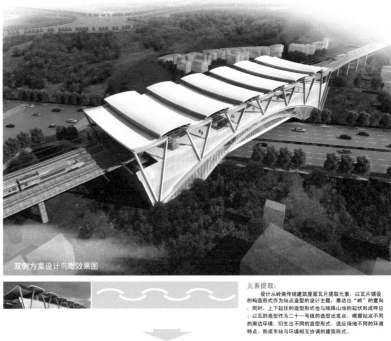

双侧方案设计鸟瞰效果图

方案设计透视效果图

建筑屋盖概念设计分析图

元素提取：
设计从岭南传统建筑屋面瓦片提取元素，以瓦片铺设的构造形式作为站点造型的设计主题，表达出"岭"的意向，同时，上下起伏的造型形式也与线路山地的起伏形成呼应。以瓦的造型作为二十一号线的造型出发点，根据站点不同的周边环境，衍生出不同的造型形式，适应场地不同的环境特点，形成车站与环境相互协调的建筑形式。

200–209	广州市轨道交通四号线 南沙客运港站装修设计	
	Guangzhou Metro Line 4 Decoration Project Design of Nansha Passenger Port Station	
210–217	广州市轨道交通十三号线一期 全线装修方案设计	
	Guangzhou Metro Line 13 Phase 1 Decoration Project Design of Entire Line	
218–221	广州市轨道交通十三号一期 新塘站装修方案设计	
	Guangzhou Metro Line 13 Phase 1 Decoration Project Design of Xintang Station	
222–223	广州市轨道交通十三号一期 鱼珠站装修方案设计	
	Guangzhou Metro Line 13 Phase 1 Decoration Project Design of Yuzhu Station	
224–233	广州市轨道交通十三号线一期 南海神庙站装修方案设计	
	Guangzhou Metro Line 13 Phase 1 Decoration Project Design of Nanhai God Temple Station	
234–235	广州市轨道交通十三号线二期 全线装修方案设计	
	Guangzhou Metro Line 13 Phase 2 Decoration Project Design of Entire Line	
236–237	广州市轨道交通四号线南延段 全线装修方案设计	
	Guangzhou Metro Line 4 South Extension Decoration Project Design of Entire Line	
238–239	广州市轨道交通二十一号线 全线装修方案设计	
	Guangzhou Metro Line 21 Decoration Project Design of Entire Line	
240–243	广州市轨道交通二十一号线 天河公园站装修方案设计	
	Guangzhou Metro Line 21 Decoration Project Design of Tianhe Park Station	
244–245	广州市轨道交通八号线 琶洲站装修方案设计	
	Guangzhou Metro Line 8 Decoration Project Design of Pazhou Station	
246–247	广州市轨道交通八号线北延段 全线装修方案设计	
	Guangzhou Metro Line 8 North Extension Decoration Project Design of Entire Line	

第三部分 车站公共区装修设计
Decoration Design of Station Public Area

248–249 广州市轨道交通八号线北延段 文化主题站装修方案设计
Guangzhou Metro Line 8 North Extension Decoration Project Design of Cultural Theme Station

250–251 广州市轨道交通十一号线 全线装修方案设计
Guangzhou Metro Line 11 Decoration Project Design of Entire Line

252–253 广州市轨道交通九号线 全线装修方案设计（投标方案）
Guangzhou Metro Line 9 Decoration Project Design of Entire Line（Bidding Proposal）

254–255 东莞地铁一号线 全线装修方案设计（投标方案）
Dongguan Metro Line 1 Decoration Project Design of Entire Line（Bidding Proposal）

256–257 青岛地铁一号线 全线装修方案设计（投标方案）
Qingdao Metro Line 1 Decoration Project Design of Entire Line（Bidding Proposal）

258–261 上海地铁迪士尼站装修方案设计（投标方案）
Shanghai Metro Decoration Project Design of Disney Resort Station（Bidding Proposal）

262–263 广州市轨道交通十四号线二期、三号线东延段、五号线东延段、八号线北延段 公共区装修方案设计
Guangzhou Metro Line 14 Phase 2, Line 3 East Extension, Line 5 East Extension & Line 8 North Extension Decoration Design of Public Area

262–263 广州市轨道交通三号线 全线装修方案设计（投标方案）
Guangzhou Metro Line 3 Decoration Project Design of Entire Line（Bidding Proposal）

262–263 广州市轨道交通十三号线 车站文化墙方案设计
Guangzhou Metro Line 13 Culture Wall Project Design

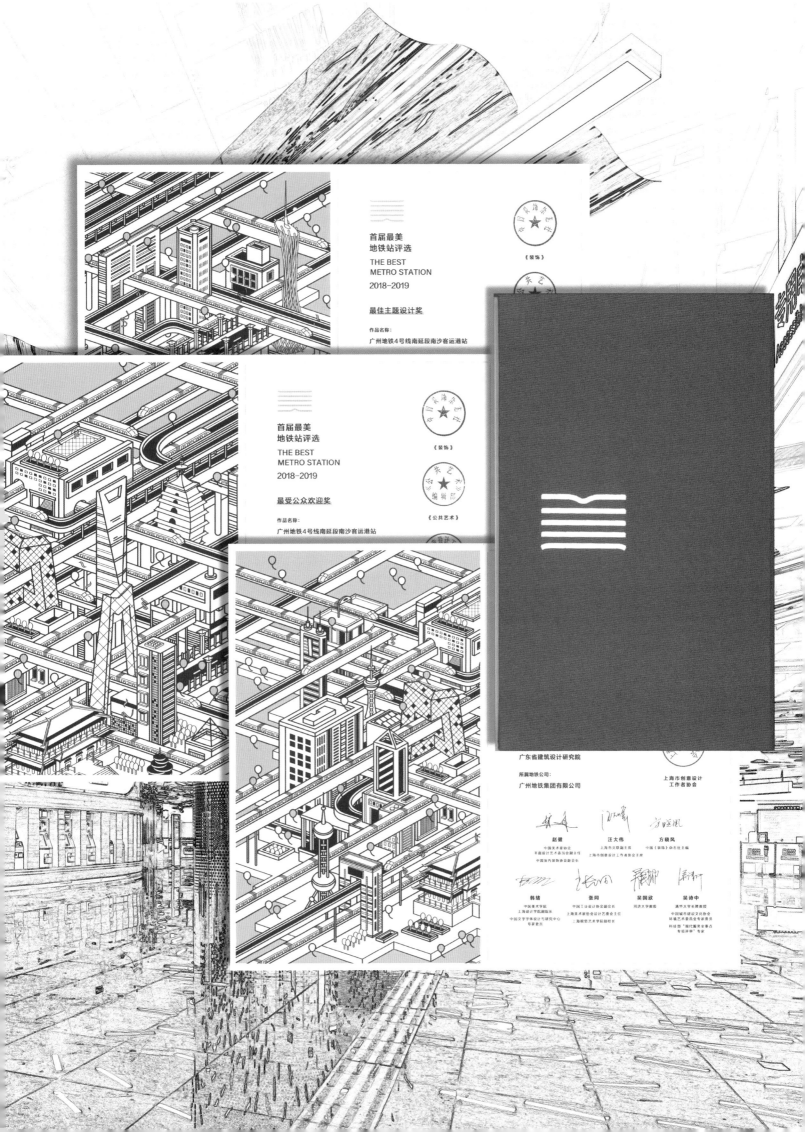

站厅空间实景

广州市轨道交通四号线
南沙客运港站装修设计
Guangzhou Metro Line 4
Decoration Project Design of Nansha Passenger Port Station

南沙客运港在粤港澳大湾区重要国家战略和"一带一路"倡议的发展背景下，其车站的文化表达需要提升到一个全新的高度，力求整体打造一个场景式的车站空间，身临其境般地体验"海上丝绸之路"的宏大。

通过蓝色喷涂土建与管线、蓝色艺术石材地面、渐变图案蓝色玻璃墙面及复合透光石材墙裙、不锈钢蚀刻图案柱子、"海鸥"灯具等元素，形成海天一色的梦幻场景。站厅中部设"宝船"艺术装置，内嵌自助售票机等服务设备，为车站空间画龙点睛，寓意南沙在全新发展中"宝船起航"。

车站公共区装修设计

空间局部实景

电梯空间局部实景

站厅室内实景

站厅室内实景

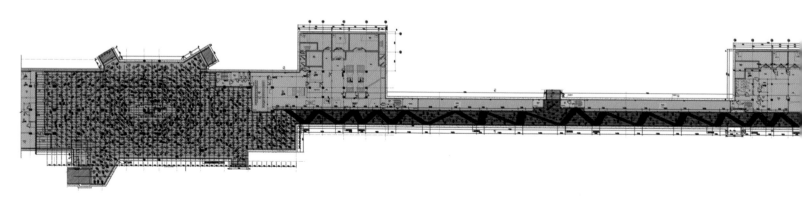

站厅及长通道吊顶平面装修详图

空间实景

站厅室内实景

站厅室内实景

转换层室内实景

转换层室内实景

转换层局部实景

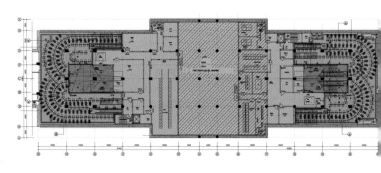
转换层吊顶平面图

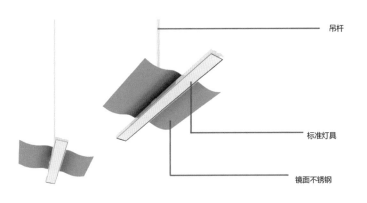

吊杆

标准灯具

镜面不锈钢

吊顶灯具的创意设计图

站台室内实景

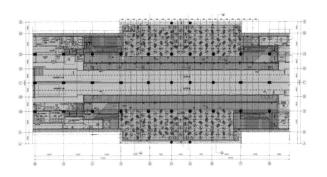

站厅层吊顶平面图

站台室内实景

207

不锈钢蚀刻图案圆柱特写实景

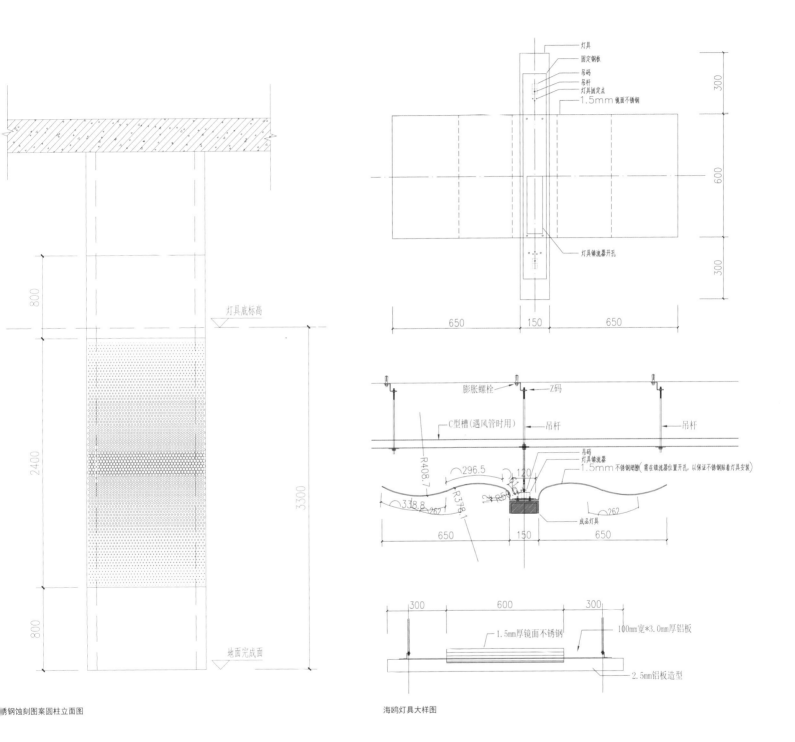

站厅内实景

广州市轨道交通十三号线一期
全线装修方案设计
Guangzhou Metro Line 13 Phase 1
Decoration Project Design of Entire Line

广州市轨道交通十三号线的车站装修追求"简捷"的现代交通理念，通过装修引导交通客流，强化交通流线与空间特征。经过多阶段方案的对比与反思，最终以"粤商珠水"作为线路文化设计概念，对模数化的弧形吊顶灯具一体式构件进行组合，形成水波图案灯带，水波流动方向与客流交通及换乘方向一致，成为"粤商珠水"中的"水流，客流，商流"，使线路文化理念和交通核心逻辑达到高度统一。

方案过程研究

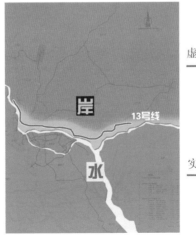
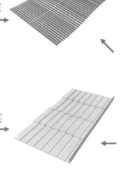
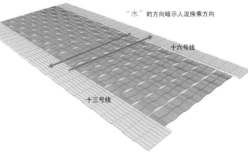

"水"的方向暗示人流换乘方向

站厅内实景

扶梯处效果

站厅内实景　　　　　　　　　　　　　　　　　　　　　站台内实景

实施方案："粤商珠水"
商流·水流·客流

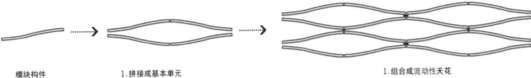

模块构件　　1.拼接成基本单元

1. 组合成流动性天花
2. 标准站：顺应客流方向
3. 换乘站：表达客流变换方向
4. 象征海上丝绸之路及碧波万顷的海面.

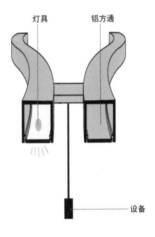

吊顶装修方案设计

站台内实景

站厅内实景

站厅内实景

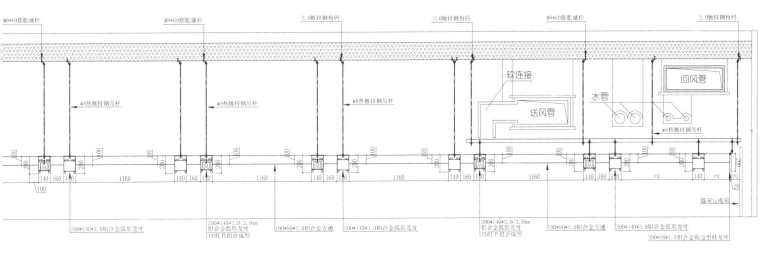

厅吊顶标准剖面图

顶局部实景

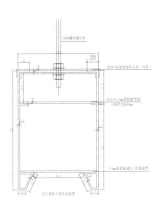
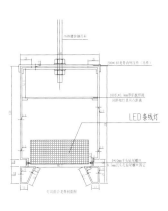

吊顶铝合金龙骨尺寸图

端厅实景

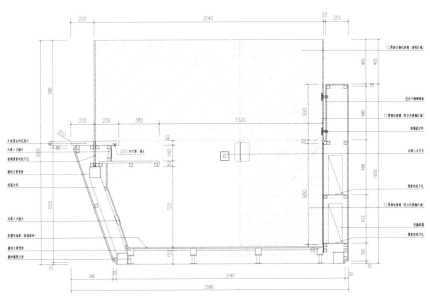

站厅票亭大样图

217

站厅室内实景

广州市轨道交通十三号一期
新塘站装修方案设计
Guangzhou Metro Line 13 Phase 1
Decoration Project Design of Xintang Station

新塘站作为十三号线与十六号线的平行换乘车站，站厅比一般车站宽敞，在室内装修上，以寻求装修元素与交通属性的内在联系为原则，通过吊顶波浪形灯带组成"水流"流动景象，水流方向与客流换乘的方向一致，通过灯光引导客流换乘。车站柱子通过弧形铝板造型与吊顶灯带衔接，形成站厅中的"树阵"，使宽敞的站厅形成独特的空间特征。

站台室内实景

车站公共区装修设计

室内实景

站台室内实景

站厅室内实景

站厅室内实景   站厅室内实景

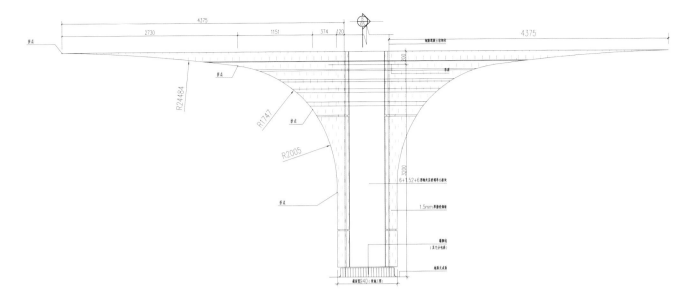

方柱立面图

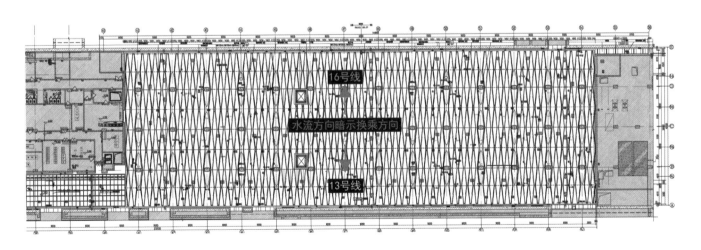

站厅层吊顶平面图

站厅室内实景

广州市轨道交通十三号一期
鱼珠站装修方案设计
Guangzhou Metro Line 13 Phase 1
Decoration Project Design of Yuzhu Station

鱼珠站为十三号线一期与既有线路五号线十字换乘的地下车站，新线路的装修风格与既有线路的装修如何协调是装修设计中需要重点考虑的问题。十三号线站厅区在吊顶系统上延续全线"粤商珠水"的弧形波浪灯具母题，水波流动方向与换乘方向一致暗示客流引导。

新旧线十字相交站厅处，通过吊顶收口细节处理过渡新旧吊顶的关系。墙面与面沿用五号线的黄色玻璃墙板和石材地面，使车站新旧区风格整体协调统一。于鱼珠站为交通换乘节点且站点所在周边为木材市场，在车站柱子设计上作出化呼应，通过穿孔搪瓷钢板形成树型图案，暗示木材文化，形成车站个性识别

站厅室内实景

站厅换乘节点

站台室内实景

车站公共区装修设计

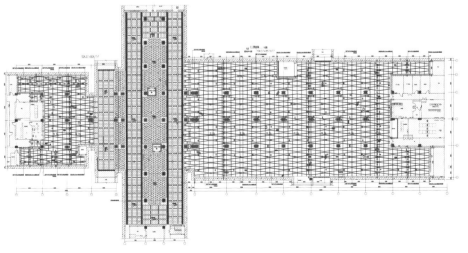

站厅吊顶平面图

站厅室内实景

广州市轨道交通十三号线一期
南海神庙站装修方案设计
Guangzhou Metro Line 13 Phase 1
Decoration Project Design of Nanhai God Temple Station

南海神庙站的装修是十三号线一期重点打造的文化主题车站，南海神庙是广州海丝文化申遗的重要部分，"海不扬波"代表着南海神庙作为出海祭祀的场所精神，车站室内装修元素通过表达"海不扬波"的海上场景，描绘出南海神庙的"海丝文化"历史画卷。在技术难点上，车站土建结构高度较为低矮，单柱站厅空间较为平淡局束，通过开放式的天花与镜面不锈钢墙面的运用，能有效打破原始空间的压抑与局束。当从通道进入车站时，地面透彻的白色石材与缓缓的红色波浪图案石材就像平静的海水在脚下静静的流淌，也象征着海丝文化源远流长。墙面不锈钢板蚀刻着从下部磨砂到上部镜面的渐变式的江崖海水纹图案，乘客仿佛置身于一望无际的大海中。不锈钢栏杆的玻璃从下到上渐变式的圆点图案表达海上雾气升腾场景，石材柱面从地上白色渐变到天空红色。开放式天花通过对管线与土建红色喷涂，进一步强化了神庙作为"海丝文化"的神圣与神秘的精神场所。天空蜿蜒曲线的不锈钢弧形灯带仿佛忆起海上和神庙中炊烟缭绕的神秘场景。雾气从地面升腾到柱子，再升腾到天空。新时代新征途，我们沿着古代海上丝绸之路，一带一路，再次启航。

车站公共区装修设计

站厅室内实景

站厅室内实景　　站台室内实景

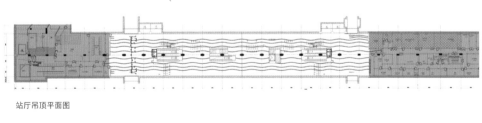

站厅吊顶平面图

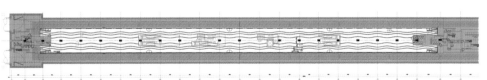

站台吊顶平面图

站台室内实景

不锈钢墙面实景图

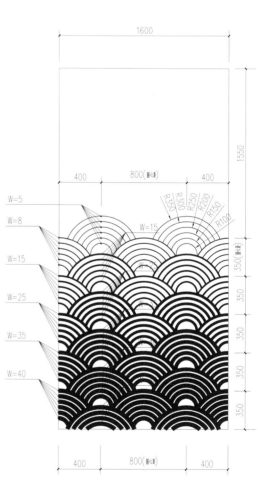

不锈钢墙面花纹详图

顶局部实景

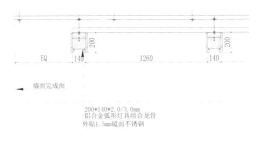

200*140*2.0/3.0mm
铝合金弧形灯具组合龙骨
外贴1.5mm镜面不锈钢

吊顶大样图

楼梯空间实景

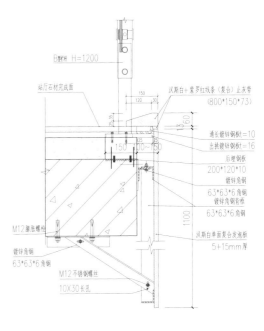

止灰带大样图

231

站厅局部实景

无柱站厅效果图

广州市轨道交通十三号线二期
全线装修方案设计
Guangzhou Metro Line 13 Phase 2
Decoration Project Design of Entire Line

广州市轨道交通十三号线二期车站装修在延续一期"粤商珠水"概念基础上，作了进一步提升。由于二期线路东西向横穿广州核心城区，属于广州城市发展的高光区域，所以装修理念应符合线路的地域特性，从"粤商珠水"中提炼出"水之高光"，吊顶上通过光的组合方式，结合直接与泛光照明，塑造出柔和变幻的水波形态，犹如水面上泛起粼粼波光。

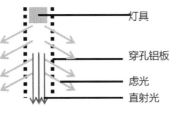

灯具
穿孔铝板
虑光
直射光

站厅照明方式概念模拟图

单柱双拱站厅效果图

多柱站厅效果图

单柱站厅效果图

车站公共区装修设计

效果图1

站厅效果图2

235

站厅室内实景

广州市轨道交通四号线南延段
全线装修方案设计
Guangzhou Metro Line 4 South Extension Decoration Project Design of Entire Line

广州市四号线南延段位于南沙区，线路从金洲延伸至南沙客运港。而南沙区作为大湾区的重要组成部分，在"一带一路"的战略发展上越显重要。车站装修以"八方来港"作为线路文化的设计概念，吊顶的AB分区延续四号线一期的设计风格，通过B区的"船桨"造型垂片和A区的穿孔铝板之间的虚实对比，体现南沙"八方来港"的"海"与"港"。

元素抽取		元素组合一
原型		
↓	→	元素组合二
抽象		
		元素组合三

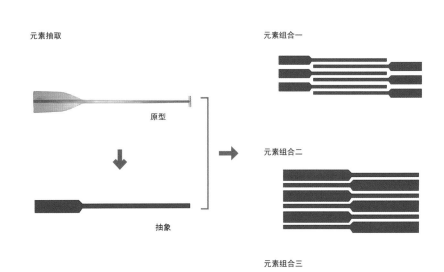

车站室内吊顶造型概念图

站台室内实景

车站公共区装修设计

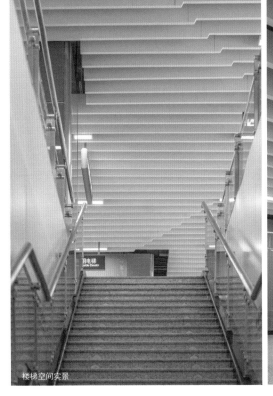

楼梯空间实景

站厅空间实景

237

站厅室内效果图

站台室内效果图

站厅室内效果图

广州市轨道交通二十一号线
全线装修方案设计
Guangzhou Metro Line 21
Decoration Project Design of Entire Line

二十一号线全线装修从传统建筑文化着手，提出传统瓦片构造形成的"岭"型，象征二十一号线途经地段的自然特色。作为车站吊顶的主要设计元素，通过波浪式的铝合金方通，打造动态的车站吊顶造型。并根据车站在两侧布置的原则，形成虚实对比的吊顶效果，满足管线遮蔽的功能。同时，在换乘车站中，根据人流导向的原则，吊顶的设计通过弧线元素，强化车站人流流线。

站厅室内效果图

站厅室内效果图

站厅

车站公共区装修设计

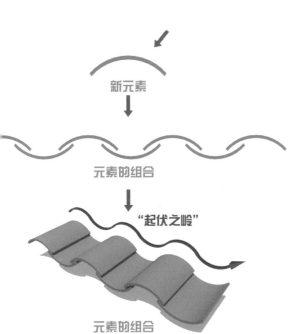

新元素

元素的组合

"起伏之岭"

元素的组合

站室内吊顶系统演变概念图

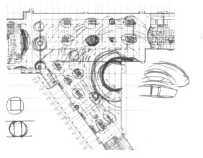 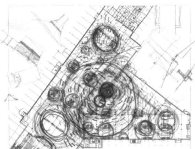 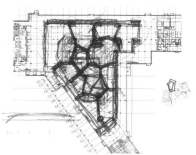 

装修方案设计构思演化图

站厅效果图

广州市轨道交通二十一号线
天河公园站装修方案设计
Guangzhou Metro Line 21
Decoration Project Design of Tianhe Park Station

天河公园为目前亚洲最大的地铁车站。车站以"浩瀚宇宙"为概念，彰显车站之大及其地域的科技文化；通过"夜空""繁星""星云""星环"等一系列元素，打造出具有一定科幻感的车站。点状灯具的排列结合车站客流分析的结果，暗示车站站厅的导向。

由于车站位于公园用地下方，车站设置了下沉庭院，形成采光中庭，为车站超大型的站厅提高了空间识别性和方向性，并给车站与公园带来了空间上及人流上的交流。

效果图

车站公共区装修设计

站厅效果图

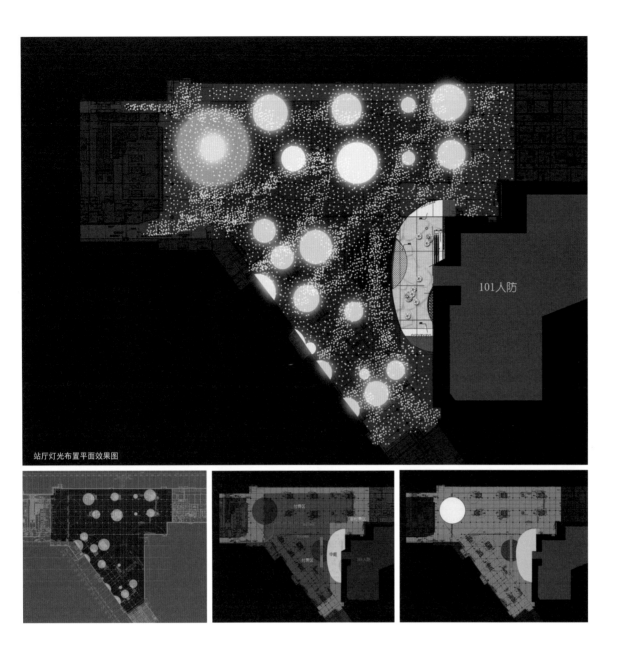

站厅灯光布置平面效果图

效果图1　　鸟瞰效果图1
效果图2　　鸟瞰效果图2

深化方案效果图

深化方案效果图

深化方案效果图

站厅室内实景

广州市轨道交通八号线
琶洲站装修方案设计
Guangzhou Metro Line 8
Decoration Project Design of Pazhou Station

八号线一期（由原二号线拆解）的装修理念是"一站一色"，不同车站墙面采用不同颜色的搪瓷钢板作为车站识别。而琶洲站墙面在选色上打破了地下站常见的思路，突破性地选用了黑色的搪瓷钢板，黑色搪瓷钢板细节上搭配不锈钢材质和鲜黄色书法站名，通过颜色与材质的对比与碰撞打造出既精致又时尚的现代交通建筑。

台室内实景

车站公共区装修设计

厅局部实景

换乘通道空间实景

站厅室内效果图

广州市轨道交通八号线北延段
全线装修方案设计
Guangzhou Metro Line 8 North Extension
Decoration Project Design of Entire Line

本线路公共区设计根据线路颜色，提出将其浓缩，形成"腰线系统"。腰线系统高199.70mm，与空间整体比例和谐美观，同时纪念广州地铁于1997年开通。腰线系统可以和地铁出入口指示、乘客方向指示、换乘指示等标识系统很好地结合，起到疏导人流的作用，同时也使空间简洁统一。吊顶设计在延续八号线元素的基础上优化，强调单一方向元素，让吊顶更加纯粹、简洁。

# 广州地铁 8号线 Guangzhou Metro Line 8

地铁开通运营时间：1997年 + 线色

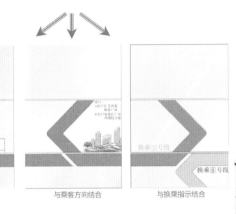

与出入口指示结合　　与乘客方向结合　　与换乘指示结合

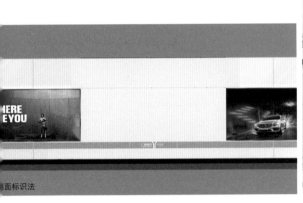

墙面标识法

站一色示意

车站公共区装修设计

站台室内效果图

人文·同德站

文旅·陈家祠站

广州市轨道交通八号线北延段
文化主题站装修方案设计
Guangzhou Metro Line 8 North Extension
Decoration Project Design of Cultural Theme Station

广州市轨道交通八号线北延段在车站装修设计上对三个车站进行了文化主题重点设计，分别为华林寺、陈家祠、同德站，从车站文化背景挖掘与提炼出共同的文化精神元素——"文"。

华林寺站属于佛教文化，车站通过"经文"元素表达佛家精神，感受佛教的"禅与空"；陈家祠站属于广州旅游名胜及氏族文化，通过不同的"陈"字表达陈家祠的文化特色；同德围片区为广州重点交通整治区域，同德站的开通能有效解决区域的交通落后问题。车站通过"同心同德"的孔明灯元素，表达市民对地铁圆梦的冀盼，体现社会"人文"精神。

站厅室内效果图

站厅室内效果图

经文 · 华林站

站厅室内效果图

站厅室内效果图

站台设计室内效果图

广州市轨道交通十一号线
全线装修方案设计
Guangzhou Metro Line 11
Decoration Project Design of Entire Line

广州市轨道交通十一号线是广州环线，线路与市中心放射型的线网交会，途经众多广州重要地带与交通节点，成为广州地铁的"广州之环"。

环状的线路把广州划分为内城外郭，装修上提取"城廓"的概念，把车站的付费区与交通核心定义为"城"，非付费区定义为"廓"。在装修上不同站点个性化表达付费区"城"的精髓，强化客流引导。弱化非付费区"廓"的装修元素，全线统一，与付费区形成对比。

付费区与非付费区的设计对比效果图

站厅室内效果图

站厅室内效果图

站厅室内效果图

广州市轨道交通九号线
全线装修方案设计（投标方案）
Guangzhou Metro Line 9
Decoration Project Design of Entire Line（Bidding Proposal）

广州市轨道交通九号线全线车站室内装修遵循"大共性，小个性，突出交通核心"的设计原则，在满足"工业化，标准化，模数化"的前提下，进一步强化车站功能空间识别和客流方向引导。

不同站点通过强化付费区装修色彩与元素，突出站点个性，引导客流进入。而全线非付费区通过统一装修风格，弱化与简化非付费区表达，突出线路共性。

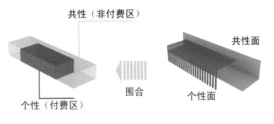

新地铁站空间

站台室内效果图

站台室内效果图

站台室内效果图

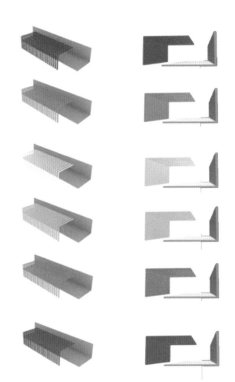

设计对比分析图

室内效果图

站台室内效果图

车站公共区装修设计

东莞地铁一号线
全线装修方案设计（投标方案）
Dongguan Metro Line 1
Decoration Project Design of Entire Line（Bidding Proposal）

东莞地铁一号线线路上经过多个自然风景区和园林等，乘客可以感受到东莞岭南的一线风光。

装修设计概念上提出了"一线山水"，通过现代的装修材料表达东莞的"山与水"。设计理念贯穿室内装修、高架站造型和出入口等，打造属于一号线专属的风格标志。

站厅室内效果图

站厅室内效果图

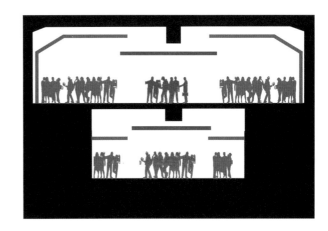

切割空间

弱化车站扁平的压抑感
吊顶高差变化暗示功能区域的变化

车站空间设计概念分析图

站厅室内效果图

地面出入口效果图
地面出入口效果图
地面出入口效果图

地面车站效果图

站厅室内效果图

青岛地铁一号线
全线装修方案设计（投标方案）
Qingdao Metro Line 1
Decoration Project Design of Entire Line（Bidding Proposal）

青岛一号线装修概念尝试打破由"文化"带来的束缚，通过现代的装修理念表达时代的青春与科技，书写青岛写意文艺的城市情怀。
注重不同车站的不同空间表达，通过色彩与灯光的运用和虚实对比，表达一个属于青岛的现代交通建筑空间。

车站公共区装修设计

站厅室内效果图

上海地铁
迪士尼站装修方案设计（投标方案）
Shanghai Metro
Decoration Project Design of Disney Resort Station（Bidding Proposal）

项目设计为中国文化与迪士尼文化寻找一个交集——"欢乐"，以中国式的欢乐和中国节日元素为主题，包括舞龙、舞师、灯笼、鞭炮、挂彩等欢乐影像传承"不折不扣中国风"，以迪士尼卡通精神融入车站的各种功能设施中，从车站造型、室内装修、灯光照明、景观园林全面唤起"原汁原味迪士尼"的情怀。

车站公共区装修设计

地面车站鸟瞰效果图

地面车站鸟瞰效果图

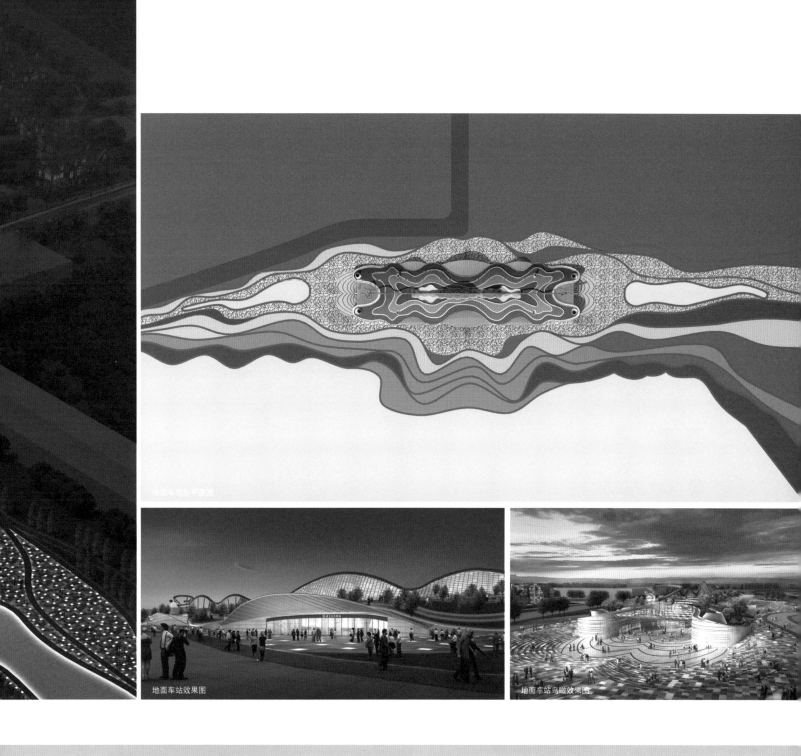

地面车站总平面图

地面车站效果图

地面车站鸟瞰效果图

广州市轨道交通十四号线二期、三号线东延段，五号线东延段，八号线北延段
公共区装修方案设计
Guangzhou Metro Line 14 Phase 2, Line 3 East Extension, Line 5 East Extension & Line 8 North Extension
Decoration Design of Public Area

广州轨道交通三号线与五号线是既有开通线路，其车站装修追求工业化、单元化、模数化的设计理念。在其线路延长段的装修设计概念中，为保持线路自身的现有特征，遵循新旧线路装修风格统一协调的原则，对既有线路的吊顶元素进行提炼和延续，通过工艺的改进、节点的优化，将单元式的模块运用得更为极致；吊顶与灯具整体结合强化特征，地面延续原线路石材，墙面使用单元式玻璃墙板系统，但在模数上延续原线路搪瓷钢板尺寸，在理性中求创新，在延续中求变化。

广州市轨道交通三号线
全线装修方案设计（投标方案）
Guangzhou Metro Line 3 Decoration Project Design of Entire Line (Bidding Proposal)

广州市轨道交通三号线车站室内装修设计遵循车站主体空间与功能的特征，在特定站点挖掘车站的地域文化并表达于装修设计元素之中。
吊顶设计中首次提出A、B分区的设计理念。A区为公共区两侧管线密集区域，吊顶形式较为密闭，可遮挡管线，吊顶形式全线统一；B区为公共区中部管线较少区域，吊顶形式较为开放，增加空间上升感。吊顶可根据不同站点设计出镂空多变的形式，成为车站个性的表达区与识别区。

广州市轨道交通十三号线
文化墙方案设计
Guangzhou Metro Line 13 Culture Wall Project Design

文化墙是车站个性识别和地域文化表达的一种重要方式，也是地铁人文情怀向大众展示的一种态度。文化墙应与车站装修整体设计元素融合，通过抽象的元素表达不同的站点文化。
广州市轨道交通十三号线设计中，文化墙不单是一个展示橱窗，而是通过更多方式与乘客互动，体现地铁开放、多元、亲民、创新的企业文化。

内效果图

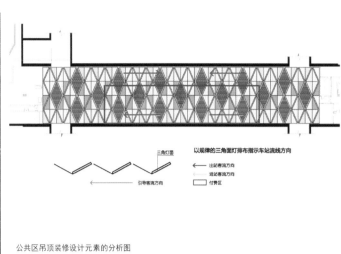
公共区吊顶装修设计元素的分析图

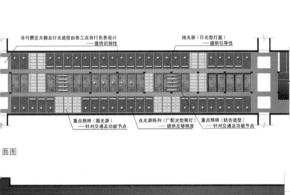
平面图

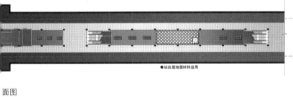
立面图

站厅室内效果图

站台室内效果图

站台室内效果图

站厅室内效果图

车站公共区装修设计

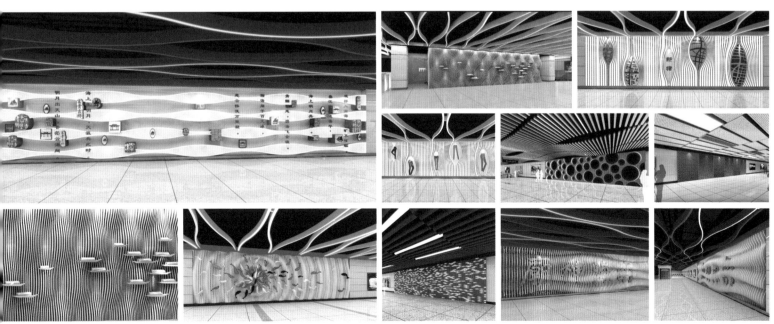

266–271	广州地铁 出入口及附属建筑设计与后期优化设计	
	Guangzhou Metro Exit & Entrance, Auxiliary Building and Further Optimization Design	
272–275	广州市轨道交通广佛线 出入口方案设计（投标方案）	
	Guangzhou Metro Guangfo Line Proposal Design of Exit & Entrance（Bidding Proposal）	
276–277	广州市轨道交通十三号线 南海神庙站出入口及附属风亭方案设计	
	Guangzhou Metro Line 13 Proposal Design of Nanhai God Temple Station Exit & Entrance and Ventilation Pavilion	
278–279	佛山地铁二号线 出入口方案设计	
	Foshan Metro Line 2 Proposal Design of Exit & Entrance	
280–281	无锡地铁一号线 出入口方案设计（投标方案）	
	Wuxi Metro Line 1 Proposal Design of Exit & Entrance（Bidding Proposal）	
282–283	广州市轨道交通十一号线出入口及附属建筑方案设计（投标）	
	Guangzhou Metro Line 11 Station Exit & Entrance and Auxiliary Building Design(Bidding)	
284–285	广州市轨道交通二号线 广州火车站地面设备用房建筑设计	
	Guangzhou Metro Line 2 Architectural Design of Guangzhou Railway Station Ground Equipment Room	
286–289	广州地铁治安监控通信指挥中心	
	Guangzhou Metro Public Security Command Center	
290–291	广州市轨道交通二十一号线 镇龙车辆段综合楼外立面及重要部位装修方案设计	
	Guangzhou Metro Line 21 Zhenlong Deport Control Center Facade Design	
292–295	广州市轨道交通二十一号线 镇龙车辆段控制中心外立面及重点空间装修方案设计	
	Guangzhou Metro Line 21 Decoration Design of Zhenlong Deport Complex Building Facade and Important Space	
296–297	广州市轨道交通二十一号线 水西停车场综合楼外立面及重点空间装修方案设计	
	Guangzhou Metro Line 21 Decoration Design of Shuixi Parking Lot Complex Building Facade and Important Space	
298–299	广州市轨道交通二十一号线 象岭停车场综合楼外立面及重点空间装修方案设计	
	Guangzhou Metro Line 21 Decoration Design of Xiangling Parking Lot Complex Building Facade and Important Space	
300–305	广州市轨道交通十三号线 官湖车辆段控制中心外立面及重点空间装修方案设计	
	Guangzhou Metro Line 13 Decoration Design of Guanhu Deport Control Center Facade and Important Space	

第四部分 轨道交通配套建筑及景观
Metro Auxiliary Building and Landscape

306–307	广州市轨道交通八号线北延段 白云湖车辆段控制中心外立面及重点空间装修方案设计
	Guangzhou Metro Line 8 North Extension Baiyun Lake Deport Control Center Facade and Important Space Proposal Design
308–309	无锡地铁 胜利门广场景观方案设计（投标方案）
	Wuxi Metro Shenglimen Square Station Landscape Design (Bidding Proposal)
310–311	广州市轨道交通五号线 动物园站站外公园大门设计
	Guangzhou Metro Line 5 Design of the Park Gate Outside Zoo Station
312–315	佛山地铁二号线 林岳西车辆段上盖开发方案设计
	Foshan Metro Line 2 Development Project Design of Building atop Linyuexi Deport
316–317	广州地铁大厦投标建筑方案设计（投标方案）
	Architectural Proposal Design of Guangzhou Metro Building (Bidding Proposal)
318–321	广州市轨道交通四号线 飞沙角车辆段上盖开发方案设计（中标方案）
	Guangzhou Metro Line 4 Development Project Design Bidding Proposal of Building atop Feishajiao Deport (Bid–Winning Proposa)
322–323	宁波地铁 车辆段上盖开发方案设计（投标方案）
	Ningbo Metro Development Project Design Bidding Proposal of Building atop Depot (Bidding Proposal)
324–325	广州市轨道交通九号线 出入口及附属建筑方案设计（投标方案）
	Guangzhou Metro Line 9 Exit & Entrance and Auxiliary Building Design (Bidding Proposal)
324–325	广州市轨道交通九号线 控制指挥中心外立面方案设计（投标方案）
	Guangzhou Metro Line 9 Decoration Design of Command and Control Center Building Facade (Bidding Proposal)
324–325	广州市轨道交通三号线 石牌桥站风亭（丰兴广场合建）
	Guangzhou Metro Line 3 Shipaiqiao Station Ventilation Pavilion (Co–constructed with Fengxing Square)
326–327	广州市轨道交通二十一号线 象—钟区间变电所
	Guangzhou Metro Line 21 Xiang–Zhong Section Substation
326–327	广州市轨道交通二十一号线 朱—象区间变电所
	Guangzhou Metro Line 21 Zhu–Xiang Section Substation
326–327	广州市轨道交通八号线 沙园站风亭（乐峰广场合建）
	Guangzhou Metro Line 8 Shayuan Station Ventilation Pavilion (Co–constructed with Lefeng Square)

"飞顶"出入口实景

广州地铁
出入口及附属建筑设计与后期优化设计
Guangzhou Metro
Exit & Entrance, Auxiliary Building and Further Optimization Design

广州市轨道交通二号线首次提出"工业化、标准化、模数化"的设计原则。在地铁出入口设计上,强调现代简洁,以交通功能为核心,通过轻盈的钢结构框架和轻薄的金属顶棚,结合通透的钢化玻璃,形成"飞顶"造型出入口,其轻柔飘逸的造型也代表着广州的城市个性。至此,广州地铁的"飞顶"出入口经过数次优化和改进一直沿用至今,成为广州地铁的重要标志。

广州市轨道交通四号线南延段属于四号线的延长线,其出入口设计主要延续四号线的原"飞顶"出入口的风格,部分"飞顶"出入口结合无障碍电梯设计,通过石材片墙的设置,无障碍电梯与"飞顶"以虚实相衬的方式结合,高风亭与低风亭通过光面与毛面石材搭配干挂而成,在功能中体现建筑的细节变化。

广州市轨道交通十三号线一期在出入口设计上,延续广州地铁"飞顶"的基本形式,细节处理上提出了更多优化。对原"飞顶"的钢结构框架和繁琐的接构件等进行了简化,其现代精致简洁的细节处理方式,体现出现代交通建筑"简·捷"与科技的属性。风亭的设计通过风口位置梳理出简洁的几何形关系,通过石材与陶棍虚实搭配,形成功能化、工业化、景观化的地铁附属建筑。

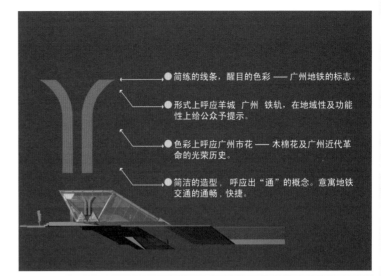

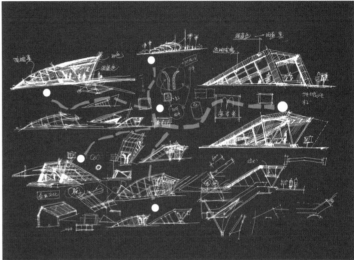

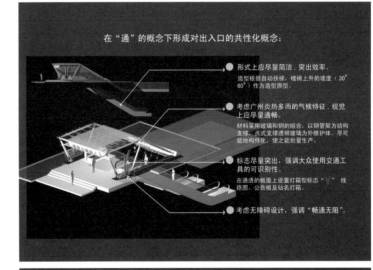

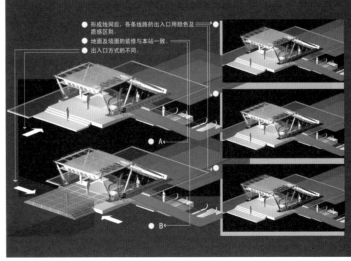

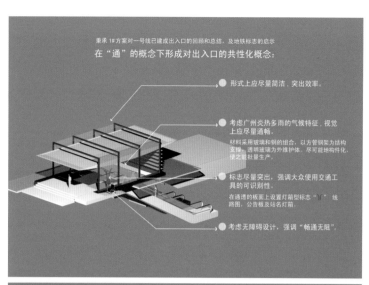

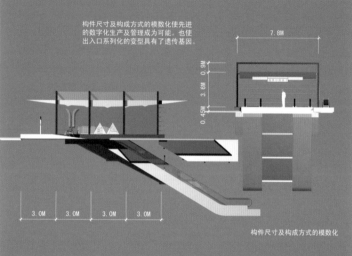

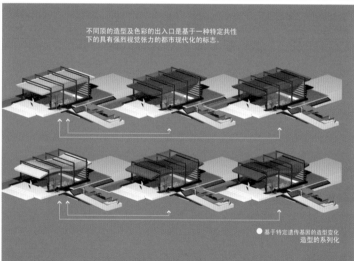

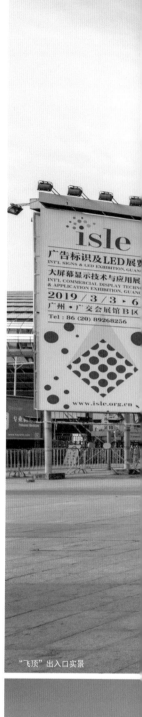

"飞顶"出入口实景

"飞顶"出入口实景

"飞顶"出入口实景

"飞顶"出入口实景

"飞顶"出入口实景

出入口鸟瞰效果图

低点效果图

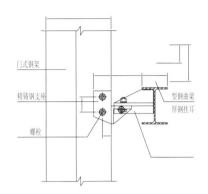
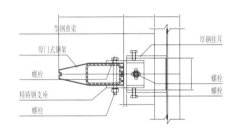
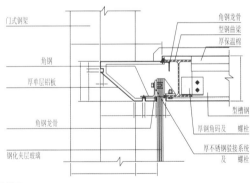

出入口建筑结构大样图

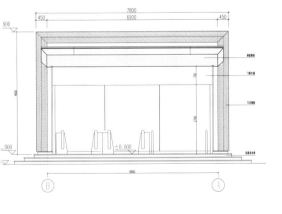
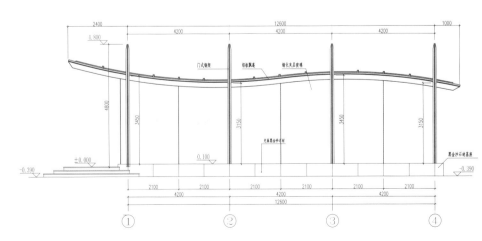

出入口建筑立面图

出入口低点效果图

出入口鸟瞰效果图

广州市轨道交通广佛线
出入口方案设计（投标方案）
Guangzhou Metro Guangfo Line
Proposal Design of Exit & Entrance（Bidding Proposal）

项目设计以简洁的钢结构及玻璃幕墙为主，虚与实两种体块结合，体现时代感和交通建筑特征。与周边城市环境相融合，具有很强的识别性。

融入历史与人文环境，继承历史、融入城市、活化地域，代表着文化的融合与发展。出入口采用模数化设计，实现工厂生产、现场装配，缩短现场安装工期，以减少对场地交通的干扰和影响。

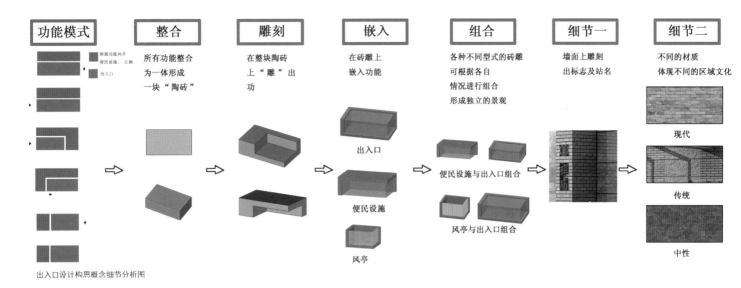

出入口设计构思概念细节分析图

出入口鸟瞰效果图

出入口低点效果图

出入口鸟瞰效果图

出入口低点效果图

出入口鸟瞰效果图

出入口低点效果图

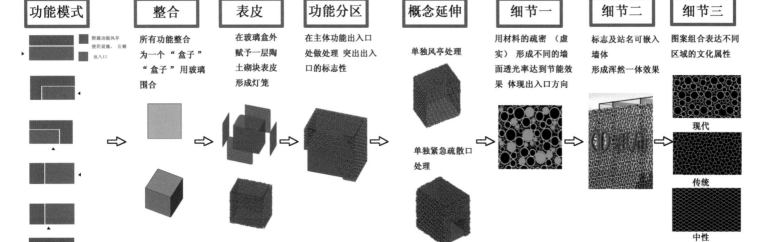
出入口设计构思概念细节分析图

低点透视效果图

鸟瞰效果图

低点效果图

出入口鸟瞰效果图

出入口鸟瞰效果图

广州市轨道交通十三号线
南海神庙站出入口及附属风亭方案设计
Guangzhou Metro Line 13
Proposal Design of Nanhai God Temple Station Exit & Entrance and Ventilation Pavilion

广州市轨道交通十三号线一期地面建筑在设计中体现现代交通建筑的"简·捷"，有效表现现代交通建筑的交通属性。

既注重交通建筑以功能为主导的设计原则，又通过整体风格的把握与个性细部的推敲，在特殊环境或历史背景如"南海神庙站"，结合地域历史特征的人文元素，使地铁设施从功能上升为地铁文化的层面。

出入口鸟瞰效果图

出入口鸟瞰效果图

口及附属风亭方案设计造型构思分析图

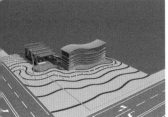 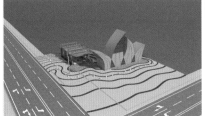 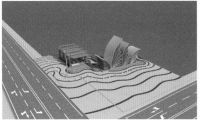 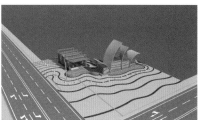

"剪影"出入口低点效果图

佛山地铁二号线
出入口方案设计
Foshan Metro Line 2
Proposal Design of Exit & Entrance

佛山二号线出入口的造型设计中，尝试通过更多代表佛山广府文化的概念表达出入口的可能性。设计在满足出入口功能的前提下，分别通过剪影、功夫、园林漏窗、岭南建筑、佛山醒狮、现代佛山等理念，形成各种形式独特的出入口景观造型，让地铁设施成为城市园林景观的亮点。

"武动禅城"出入口低点效果图

"禅城之窗"出入口方案低点效果图

"印象—佛山"出入口方案低点效果图

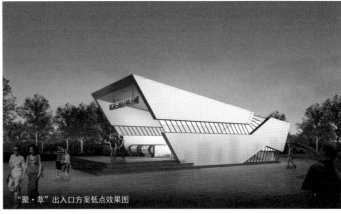
"聚·萃"出入口方案低点效果图

"武动禅城"出入口低点效果图2

"印象—佛山"出入口鸟瞰效果图

"禅城之窗"出入口鸟瞰效果图

"聚·萃"出入口鸟瞰效果图

"简·韵"出入口鸟瞰效果图

"剪影"出入口鸟瞰效果图

"武动禅城"出入口鸟瞰效果图

"简·韵"出入口低点效果图

标准出入口低点效果图

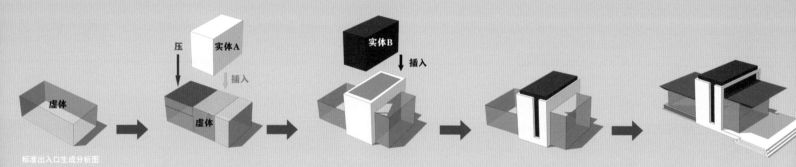

标准出入口生成分析图

无锡地铁一号线

出入口方案设计（投标方案）

Wuxi Metro Line 1
Proposal Design of Exit & Entrance（Bidding Proposal）

无锡作为著名的江南文化名城，在一号线高架站的造型设计中，吸收江南文化的精髓，把江南文化与地铁交通功能紧密结合，设计中提出"一式异景"的设计原则，用一种地域风格和构成模式，通过不同的文化背景打造一条线路中的不同车站。
车站充分考虑当地的气候特征及本土产业无锡光伏玻璃板的利用，形成独特的"无锡模式"地铁文化标杆。

标准出入口低点效果图

标准出入口鸟瞰效果图

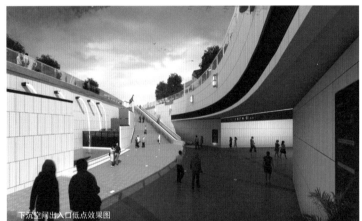

下沉空间出入口低点效果图

下沉空间出入口鸟瞰效果图

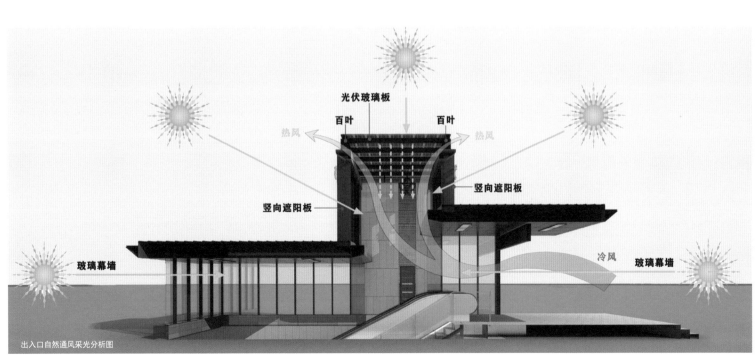

出入口自然通风采光分析图

附属建筑低点效果图

出入口低点效果图

出入口低点效果图

广州市轨道交通十一号线
出入口及附属建筑方案设计（投标）
Guangzhou Metro Line 11
Station Exit & Entrance and Auxiliary Building Design(Bidding)

广州市轨道交通十一号线出入口在既有线路出入口基础上，提出新的设计。在维持弧形顶棚设计的同时，根据功能与体量，通过局部的实墙，与出入口玻璃围护结构形成虚实对比，丰富建筑立面。

风亭结合垂直绿化，通过横向的花槽整合绿化，并成为风亭立面的主要元素，利用花槽错动的变化，使风亭立面产生一定特征性。

出入口低点效果图

风亭低点效果图

附属建筑鸟瞰效果图

建筑实景

广州市轨道交通二号线
广州火车站地面设备用房建筑设计
Guangzhou Metro Line 2
Architectural Design of Guangzhou Railway Station Ground Equipment Room

广州市轨道交通二号线广州火车站地面设备用房位于广州火车站广场上，周边人流密集。

设备用房造型采用明确的体块关系，通过不同颜色的体量穿插，形成细致、丰富的立面关系，形成雕塑般的室外造型，提高了广场景观的丰富性。

建筑局部实景

轨道交通配套建筑及景观

建筑南面低点实景

广州地铁
治安监控通信指挥中心
Guangzhou Metro
Public Security Command Center

广州地铁公安监控通信指挥中心项目位于广州市轨道交通赤沙车辆段，总建筑面积1.14万m²。针对治安监控通信指挥的功能属性，设计以"石""木"的理念，体现"严格执法、热情服务"的精神，通过石材和木材在不同场所的穿插运用，塑造建筑表情，使建筑室内外设计元素高度一致，外墙细节元素追求尽可能完美精致，建筑在朴实中体现气质与品位。

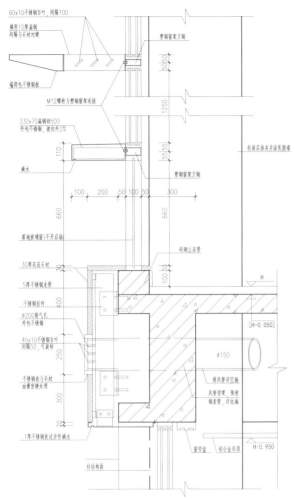

墙及窗户大样图

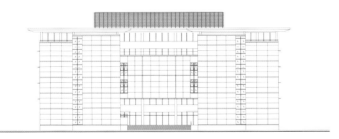

立面图

立面图

建筑外墙及窗户局部实景

建筑局部实景

建筑低点实景

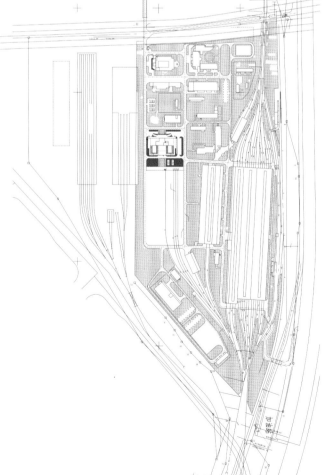

总平面图

二层平面图

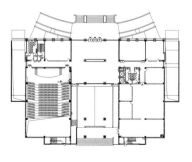

首层平面图

建筑北面低点实景

建筑正面出入口局部实景

建筑低点实景

建筑空间局部实景

建筑正面出入口局部实景

建筑低点实景

广州市轨道交通二十一号线
镇龙车辆段综合楼外立面及重要部位装修方案设计
Guangzhou Metro Line 21
Zhenlong Deport Control Center Facade Design

镇龙车辆段综合楼受站点设置与站点周边环境条件的制约，车站站厅设于地面，组合了其他城市公用设施（如公共卫生间、变电所、社区居委会、商铺、设备房等），形成功能高度浓缩的小型城市交通综合体。建筑主体随其周边城市环境呈现自我个性的建筑表达，注重现代简洁的体量组合与穿插，并通过材料搭配对建筑细节精心刻画，打造属于新时代的交通建筑品位。

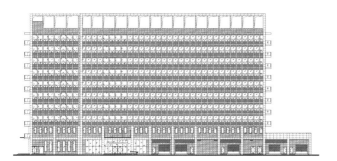
正立面图

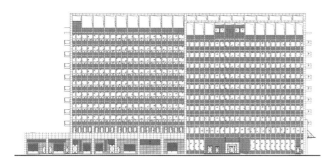
背立面图

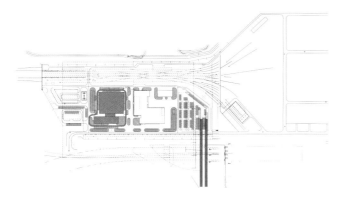
总平面图

建筑局部实景

建筑低点实景

建筑低点实景

轨道交通配套建筑及景观

建筑低点实景

广州市轨道交通二十一号线
镇龙车辆段控制中心外立面及重点空间装修方案设计
Guangzhou Metro Line 21
Decoration Design of Zhenlong Deport Complex Building Facade and Important Space

镇龙车辆段控制中心为广州地铁线网区域性的控制中心,由塔楼及裙楼监控大厅组成,塔楼建筑立面采用简洁铝挡板及大面积的石材,并通过变化的节奏塑造地铁办公建筑端庄活力的建筑形象。

裙房的立面设计中,利用体量切割的设计手法,将裙房的大体量进行细分,采用黑白灰三种颜色的石材搭配,利用错动随机的排列,形成数码化的建筑立面,暗示出控制中心的信息功能。

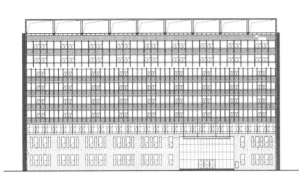
塔楼立面图

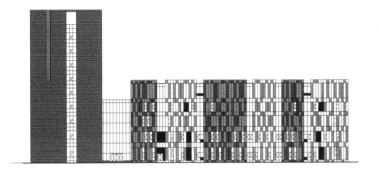
立面图

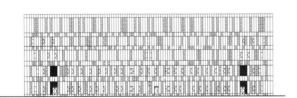
裙楼立面图

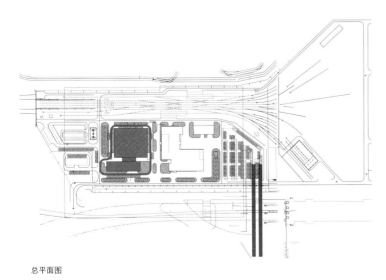
总平面图

建筑低点实景

建筑低点实景

建筑局部实景

轨道交通配套建筑及景观

293

建筑低点实景

建筑室内实景

建筑局部实景

建筑局部实景

局部实景

建筑局部实景

建筑低点实景

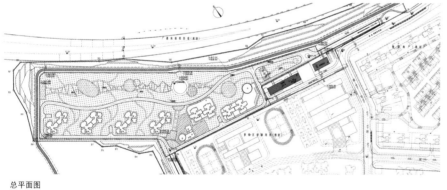

总平面图

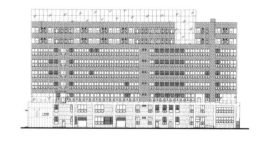

立面图

广州市轨道交通二十一号线
水西停车场综合楼外立面及重点空间装修方案设计
Guangzhou Metro Line 21
Decoration Design of Shuixi Parking Lot Complex Building Facade and Important Space

水西停车场位于广州市黄埔区萝岗区域，综合楼为9层建筑，下部为办公、餐厅及大堂等，上部为宿舍。底层的库房部分，利用连续的柱廊进行整合，避免了大小不一的库房洞口对外立面的不利影响。建筑外立面通过垂直的立面元素，将办公部分与宿舍部分的外立面有机统一起来，形成整体。整体建筑外立面采用加减法及穿插体量的手法，具有简洁现代的气质。

建筑低点实景

建筑外墙局部实景

局部实景

建筑局部实景

轨道交通配套建筑及景观

建筑低点实景

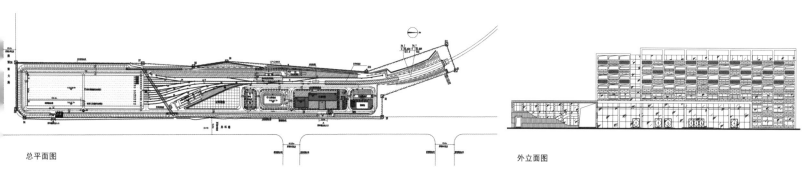

总平面图　　　　　　　　　　　　　　　　　　　外立面图

广州市轨道交通二十一号线
象岭停车场综合楼外立面及重点空间装修方案设计
Guangzhou Metro Line 21
Decoration Design of Xiangling Parking Lot Complex Building Facade and Important Space

象岭停车场位于广州市增城区，综合楼为6层建筑，功能上综合了办公、餐厅、会议、宿舍等功能。建筑针对上部宿舍功能及下部办公功能截然不同的功能形式，采用了具有丰富变化的矩形立面单元，从而将上下两种功能立面统一起来。同时，结合立面构造，在立面上营造出绿化条件，在丰富建筑立面的同时，为内部空间提供了优质的环境条件。

建筑低点实景

建筑低点实景

建筑低点实景

建筑低点实景

轨道交通配套建筑及景观

建筑鸟瞰实景

广州市轨道交通十三号线
官湖车辆段控制中心外立面及重点空间装修方案设计
Guangzhou Metro Line 13
Decoration Design of Guanhu Deport Control Center Facade and Important Space

官湖车辆段综合楼作为一个广州地铁后勤大脑的重要功能性建筑，在建筑外观造型设计中，以功能作为切入点，根据不同功能房间对通风采光的需求，设置不同的外立面开窗形式，暗示建筑中不同区域的功能属性，自然形成既独特又简洁现代的交通办公建筑。

建筑低点实景

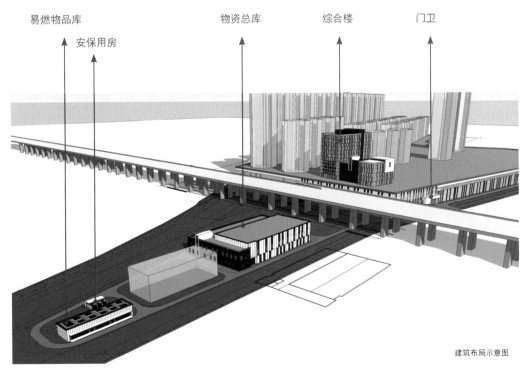
建筑布局示意图

建筑低点实景

建筑低点实景

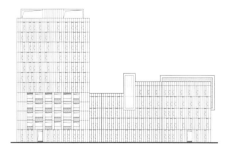

立面图1

立面图2

建筑局部实景

建筑局部实景

立面图3

立面图4

方案过程研究

鸟瞰效果图

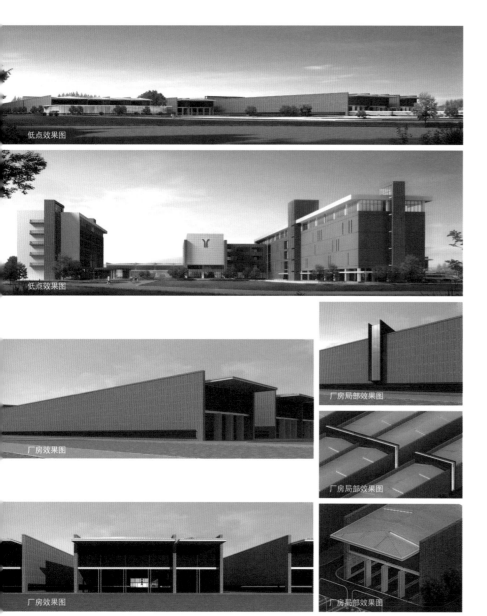
低点效果图

低点效果图

厂房效果图

厂房效果图

厂房局部效果图

厂房局部效果图

厂房局部效果图

总平面布局图

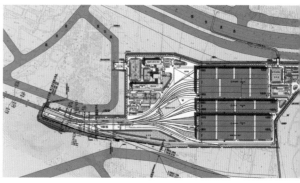
总平面图

综合楼鸟瞰效果图

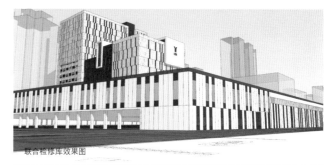
联合检修库效果图

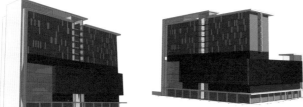

易燃物品库效果图

综合楼造型分析图

物资总库效果图

综合楼室内效果图

综合楼室内效果图

综合楼室内效果图

鸟瞰效果图

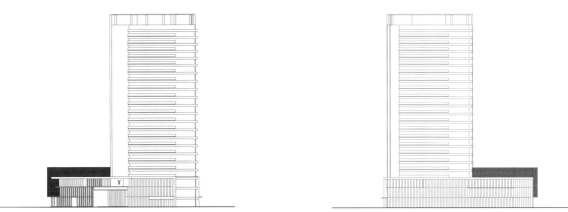

立面图1　　　　　　　　　　　　立面图2

广州市轨道交通八号线北延段
白云湖车辆段控制中心外立面及重点空间装修方案设计
Guangzhou Metro Line 8 North Extension
Baiyun Lake Deport Control Center Facade and Important Space Proposal Design

白云湖车辆段控制中心外观造型通过统一的"线条"母题，以不同的线条组合表达交通建筑的简捷现代特性，强调建筑不同的功能特征。
裙房以办公为主，外立面为竖线条，突出交通属性；塔楼为公寓宿舍，外立面为横线条，强调居住属性。项目打造了既具功能性，又极具个性的交通建筑。

大门效果图

室内电梯间效果图

大堂空间效果图

轨道交通配套建筑及景观

附属建筑效果图

车库出入口效果图

鸟瞰效果图

下沉空间低点效果图

地铁出入口低点效果图

无锡地铁
胜利门广场景观方案设计（投标方案）
Wuxi Metro
Shenglimen Square Station Landscape Design (Bidding Proposal)

无锡胜利门广场为无锡地铁胜利门站所连接的地下空间上盖广场，同时也是无锡具有革命纪念意义的城市广场。周边是高楼林立的现代城市空间，其景观设计考虑营造出兼顾大小尺度的城市休闲广场的氛围，提出"星星之火可以燎原"的设计理念，将不利的疏散楼梯间转化为景观中的重要元素，并通过植入景观天桥、城市纪念门、景观大道等空间元素，强化胜利门广场的文化背景。

鸟瞰效果图

低点效果图

总平面图

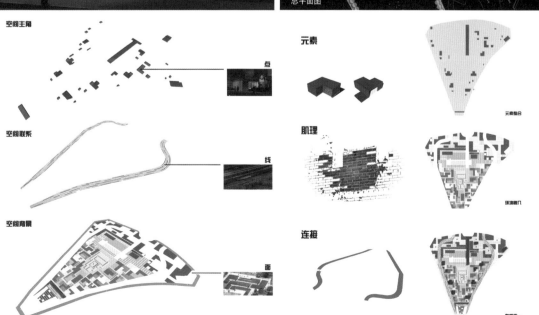

设计理念分析图

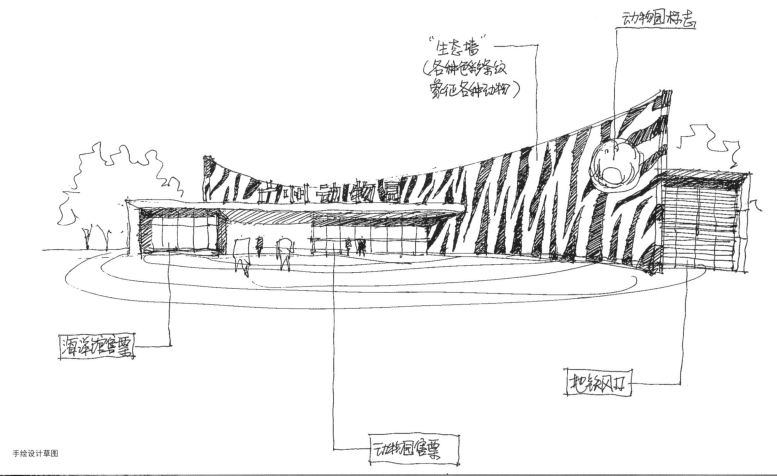

手绘设计草图

低点效果图

广州市轨道交通五号线
动物园站站外公园大门设计
Guangzhou Metro Line 5
Design of the Park Gate Outside Zoo Station

广州市轨道交通五号线动物公园站是一个充满特色的地铁车站,车站内粗犷的清水混凝土与Y字柱结构美学演绎出动物公园的自然原始气质。根据功能需求,动物公园前广场需设置车站的出入口与地面附属。

设计考虑通过整合的思想,化繁为简,将不利与矛盾转化为有利因素,对广场景观、大门形象、公园管理用房、地铁设施以集约与整合的思想,作景观化处理,形成一个形象鲜明的城市特色公园广场空间,突出动物园在城市界面中的特殊形象。

低点效果图

低点效果图

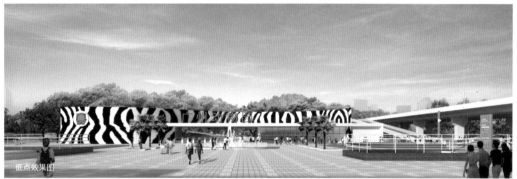
低点效果图

低点效果图

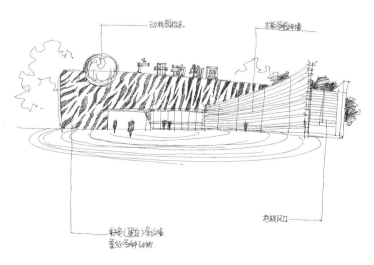
手绘设计草图

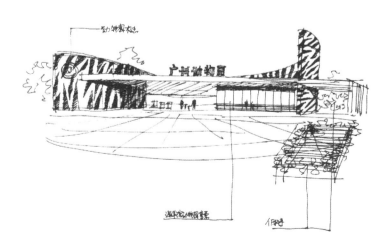
手绘设计草图

鸟瞰效果图

鸟瞰效果图

佛山地铁二号线

林岳西车辆段上盖开发方案设计
Foshan Metro Line 2
Development Project Design of Building atop Linyuexi Deport

林岳西上盖开发总用地面积67万m²，总建筑面积266万m²，方案锐意打造一个集居住、交通、商业、教育、产业于一体的艺术型高端创意综合社区。居住以"创意峡谷"为主轴，配以艺术塔、创意大楼、创意街、艺术学校、时尚天幕及酒店等，让年轻而富有创意的社交生活能在此得到无尽的创造与释放。

鸟瞰效果图

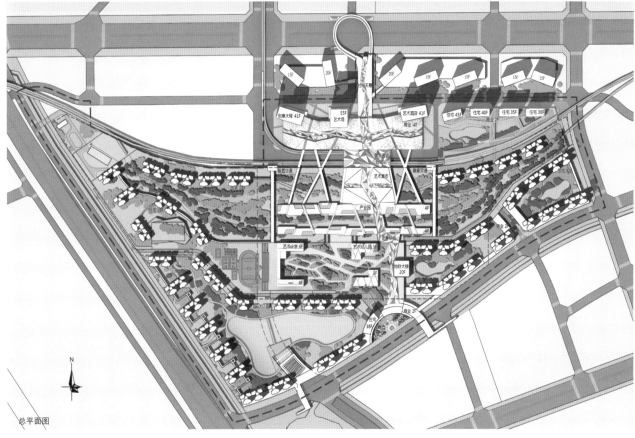
总平面图

轨道交通配套建筑及景观

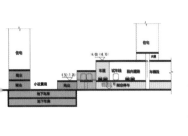
概念设计分析图

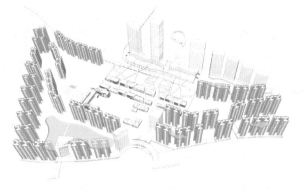

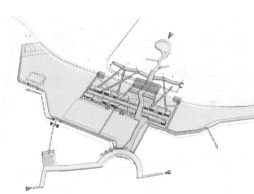

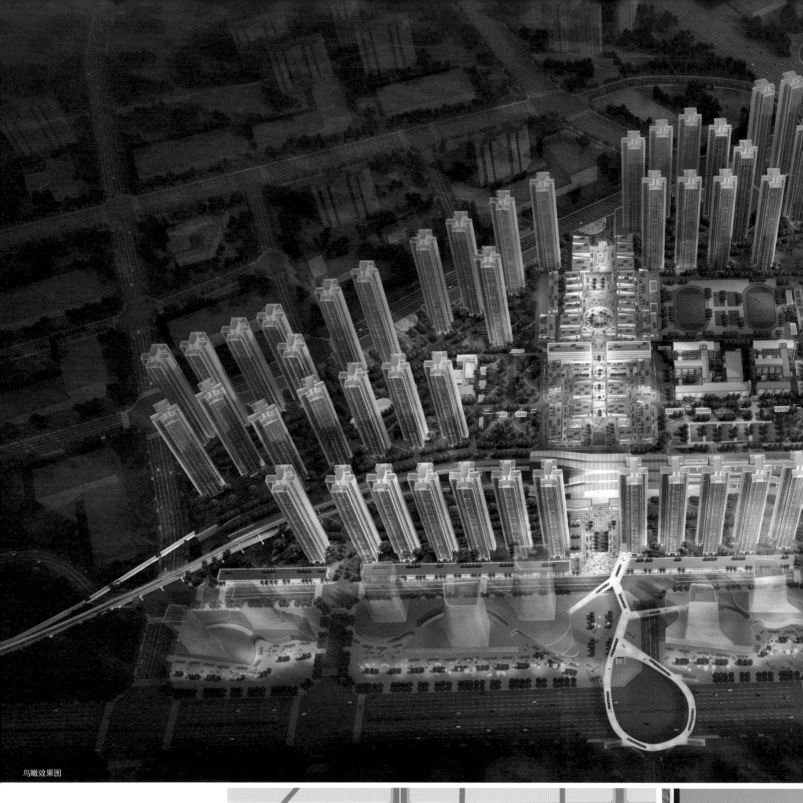
鸟瞰效果图

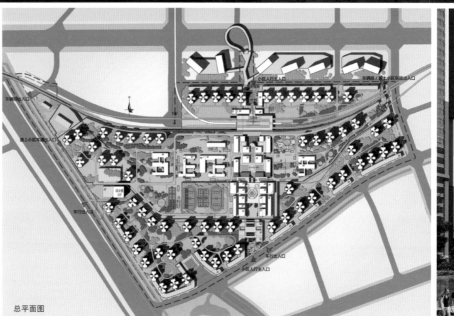
总平面图

焦点效果图

鸟瞰效果图

广州地铁大厦投标建筑方案设计（投标方案）
Architectural Proposal Design of Guangzhou Metro Building (Bidding Proposal)

广州地铁大厦设计项目为广州地铁的标志性总部大楼，设3栋塔楼与1座地铁科普中心。设计以强烈的"线性"表达建筑物的交通线性文化，突显地铁交通无限延伸与四通八达精神。

设计取消传统的裙房式商业集中布置格局，以"廊"的方式组织连接各塔楼，将休闲式商业融在"廊"道中，并与下沉广场形成立体多层次商业布局空间。

低点效果图

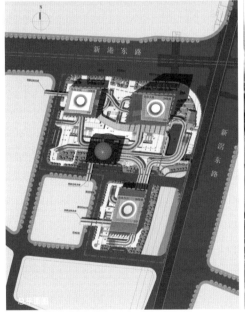

鸟瞰效果图

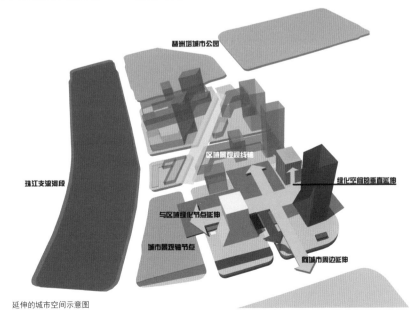

延伸的城市空间示意图

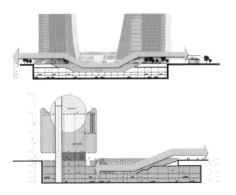

延伸的城市空间

项目设计中通过降低4#塔楼的策略，使原城市设计位于项目西侧及南侧的城市广场空间得以有效延伸，并通过地块的下沉广场使空间向下延伸。同时沿下沉广场，设计中将南地块南面城市绿地向北延伸，并沿建筑外墙向上延伸形成1#~3#塔楼的空中室外平台，南地块东面的河涌景观也延伸至北地块内，通过假山，水池，枯山水河道吗，水墙最后延伸至下沉广场中心表演舞台形成连贯的空间体验。

鸟瞰效果图

广州市轨道交通四号线
飞沙角车辆段上盖开发方案设计（中标方案）
Guangzhou Metro Line 4
Development Project Design Bidding Proposal of Building atop Feishajiao Deport (Bid-Winning Proposal)

本案规划设计在尊重用地原状的情况下，以低密度、良好的景观通达性以及多元化的商业活力空间为设计指导思想，利用台地景观对场地高差进行分解，营造趣味性的景观空间，同时激活商业氛围，创造一个人与自然和谐共生的乐土，打造一个现代都市生活理想的模式，营造一个男女老少健康生活的天堂，塑造一个形象标志性的现代居住社区。

商业广场低点效果图

主入口低点效果图

主入口低点效果图

总平面图

轨道交通配套建筑及景观

鸟瞰效果图

商业街低点效果图

住宅景观效果图

住宅景观效果图

台地景观效果图

中心园林鸟瞰效果图

幼儿园鸟瞰效果图

北侧整体低点效果图

南侧整体低点效果图

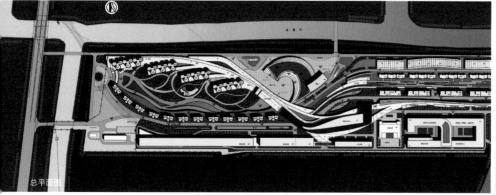

总平面图

鸟瞰效果图

宁波地铁
车辆段上盖开发方案设计（投标方案）
Ningbo Metro
Development Project Design Bidding Proposal of Building atop Depot (Bidding Proposal)

项目通过对宁波地域文化的深度挖掘，从水的动态引申出场地整体空间的形态，形成"行云流水，文商荟萃"的格局，打造集办公、住宅、休闲、交通设施、交通后勤等于一体的交通枢纽，实现土地利用集约化。
整体以车站为中心，向心式布局设计，可进一步强化TOD的理念，发挥出TOD模式的优势。

商业中心低点效果图

中心景观区低点效果图

综合楼低点效果图

公安楼低点效果图

住宅区低点效果图

轨道交通配套建筑及景观

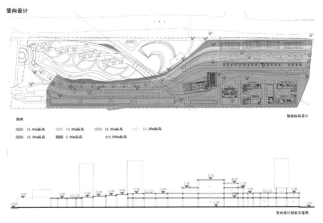
概念设计分析图

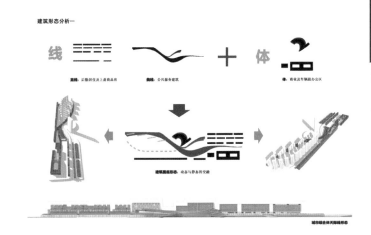

广州市轨道交通九号线
出入口及附属建筑方案设计（投标方案）
Guangzhou Metro Line 9
Exit & Entrance and Auxiliary Building Design (Bidding Proposal)

项目设计以简洁的钢结构及玻璃幕墙为主，虚与实两种体块相结合，体现时代感和交通建筑特征，与周边城市环境相融合，具有很强的识别性。融入历史与人文环境，继承历史、融入城市、活化地域，代表着文化的融合与发展。
出入口采用模数化设计，实现工厂生产、现场装配，缩短现场安装工期，以减少对场地交通的干扰和影响。

广州市轨道交通九号线
控制指挥中心外立面方案设计（投标方案）
Guangzhou Metro Line 9
Decoration Design of Command and Control Center Building Facade (Bidding Proposal)

控制指挥中心建筑景观、造型设计将传统建筑文化与现代技术、结构、材料结合起来，反映时代特点。
运用简洁的建筑语言，在传统的基础上创新，在满足使用功能前提下，体现建筑自身的性格，造型设计美观大方，景观设计舒适宜人，控制中心作为地铁轨道交通配套建筑，体现了市政公共建筑的特点，也体现了地铁文化。

广州市轨道交通三号线
石牌桥站风亭（丰兴广场合建）
Guangzhou Metro Line 3
Shipaiqiao Station Ventilation Pavilion (Co-constructed with Fengxing Square)

石牌桥站风亭位于广州市天河区丰兴广场首层建筑范围内。在设计丰兴广场的过程中，将石牌桥风亭与丰兴广场首层进行整合，通过建筑立面处理，将风亭融入建筑内，减少风亭对周边人行道空间的占用。

效果图

低点效果图

夜景效果图

低点效果图

低点效果图

轨道交通配套建筑及景观

广州市轨道交通二十一号线
象—钟区间变电所
Guangzhou Metro Line 21
Xiang-Zhong Section Substation

象—钟区间变电所通过外部简洁、动态的建筑造型，表达变电所建筑的力量与动感。建筑内部为3层，为区间动力设备提供空间载体。建筑通过立面斜向与横向错动的线条元素，形成丰富的建筑形态。深浅相间的立面横向线条，为立面提供细节，并隐喻建筑的动力、速度的属性。

广州市轨道交通二十一号线
朱—象区间变电所
Guangzhou Metro Line 21
Zhu-Xiang Section Substation

朱—象区间变电所建筑内部功能主要为变电设备及管廊，功能明确单一。根据建筑功能，采用明确的建筑体量关系，形成大实大虚的立面对比。并通过简洁的立面，呼应建筑功能，同时与周边山体环境形成明确的对比。建筑形体采用线面结合的处理方式，让建筑立面有更丰富的元素。

广州市轨道交通八号线
沙园站风亭（乐峰广场合建）
Guangzhou Metro Line 8
Shayuan Station Ventilation Pavilion (Co-constructed with Lefeng Square)

沙园站风亭位于广州乐峰广场首层建筑范围内。乐丰广场在设计过程中积极整合地铁风亭，将风亭与商场首层有机结合，满足了风亭的使用功能，并与商业形态融为一体，减少地铁风亭对城市环境的影响。

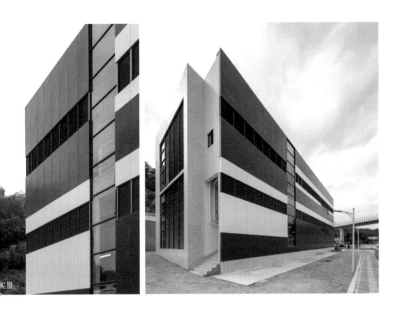

实景

低点实景

轨道交通配套建筑及景观

330–331	广州海珠环岛新型有轨电车试验段　车站方案研究——岭南画卷	
	Guangzhou Haizhu Roundabout Tram Test Section Station Project Research － Lingnan Picture Scroll	
332–333	广州海珠环岛新型有轨电车试验段　车站方案研究——龙舞岭南	
	Guangzhou Haizhu Roundabout Tram Test SectionStation Project Research － Dragon Dance in Lingnan	
334–335	广州海珠环岛新型有轨电车试验段　车站方案研究——广府趟栊	
	Guangzhou Haizhu Roundabout Tram Test SectionStation Project Research － Wooden Sliding Door of Cantonese Characteristic	
336–337	广州海珠环岛新型有轨电车试验段　车站方案研究——山岭白云	
	Guangzhou Haizhu Roundabout Tram Test SectionStation Project Research － Clouds on the Mountains	
338–339	广州海珠环岛新型有轨电车试验段　车站方案研究——广府之门一	
	Guangzhou Haizhu Roundabout Tram Test Section Station Project Research － Gate of Guangzhou 1	
340–341	广州海珠环岛新型有轨电车试验段　车站方案研究——广府之门二	
	Guangzhou Haizhu Roundabout Tram Test Section Station Project Research － Gate of Guangzhou 2	
342–343	广州海珠环岛新型有轨电车试验段　车站方案研究——广府之门三	
	Guangzhou Haizhu Roundabout Tram Test Section Station Project Research － Gate of Guangzhou 3	
344–345	广州海珠环岛新型有轨电车试验段　车站方案研究——云山珠水	
	Guangzhou Haizhu Roundabout Tram Test Section Station Project Research － Baiyun Mountain & Pearl River	
346–347	广州海珠环岛新型有轨电车试验段　车站方案研究——城市脉搏	
	Guangzhou Haizhu Roundabout Tram Test Section Station Project Research － Urban Pulse	
348–349	广州海珠环岛新型有轨电车试验段　车站方案研究——粤章	
	Guangzhou Haizhu Roundabout Tram Test Section Station Project Research － Guangzhou Chapter	
350–351	广州海珠环岛新型有轨电车试验段　车站方案研究——流动景点一	
	Guangzhou Haizhu Roundabout Tram Test SectionStation Project Research － Mobile Scenic Spot 1	
352–353	广州海珠环岛新型有轨电车试验段　车站方案研究——流动景点二	
	Guangzhou Haizhu Roundabout Tram Test SectionStation Project Research － Mobile Scenic Spot 2	
354–355	广州海珠环岛新型有轨电车试验段　车站方案研究——YOUNGTRAM 1	
	Guangzhou Haizhu Roundabout Tram Test SectionStation Project Research － YOUNGTRAM 1	

第五部分　广州海珠环岛新型有轨电车试验段
Guangzhou Haizhu Roundabout Tram Test Section

356–357	广州海珠环岛新型有轨电车试验段车站方案研究——YOUNGTRAM 2
	Guangzhou Haizhu Roundabout Tram Test Section Station Project Research YOUNGTRAM 2
358–359	广州海珠环岛新型有轨电车试验段有轨电车车站附属充电站方案研究
	Guangzhou Haizhu Roundabout Tram Test Section Auxiliary Charging Station Project Research
360–361	广州海珠环岛新型有轨电车试验段广州塔站充电站与折返线结合方案
	Guangzhou Haizhu Roundabout Tram Test Section Canton Tower Station Integration Plan of Charging Station and Turn — back Line
362–367	广州海珠环岛新型有轨电车试验段有轨电车车站实施方案设计
	Guangzhou Haizhu Roundabout Tram Test Section Station Implementation Project Design
368–369	广州海珠环岛新型有轨电车试验段有轨电车建筑泛光照明设计
	Guangzhou Haizhu Roundabout Tram Test Section Building Floodlighting Design
370–371	广州海珠环岛新型有轨电车试验段琶醍站改造设计
	Guangzhou Haizhu Roundabout Tram Test Section Pati Station Renovation Design
372–373	广州海珠环岛新型有轨电车试验段琶醍站配套附属建筑设计
	Guangzhou Haizhu Roundabout Tram Test Section Pati Station Auxiliary Building Design
374–385	广州海珠环岛新型有轨电车试验段琶醍区域立体综合交通枢纽方案研究
	Guangzhou Haizhu Roundabout Tram Test Section Pati Area Multimodal Transport Hub Project Research
386–387	广州海珠环岛新型有轨电车试验段有轨电车车辆段方案研究
	Guangzhou Haizhu Roundabout Tram Test Section Tram Deport Project Research
388–389	广州海珠环岛新型有轨电车附属配套公共卫生间
	Guangzhou Haizhu Roundabout Tram Ancillary Facility Overpass Project Design
388–389	广州海珠环岛新型有轨电车附属配套琶醍中心区改造
	Guangzhou Haizhu Roundabout Tram Ancillary Facility Central Pati Renovaton Design
388–389	广州海珠环岛新型有轨电车附属配套天桥方案设计
	Guangzhou Haizhu Roundabout Tram Ancillary Facility Overpass Project Design
390–395	广州海珠环岛新型有轨电车试验段广州塔周边结合车站配套设施环境优化设计
	Guangzhou Haizhu Roundabout Tram Test Section Environmental Optimization Design of Canton TowerPeripheral Area Combing with Station Ancillary Facility

车站鸟瞰效果图

广州海珠环岛新型有轨电车试验段
车站方案研究——岭南画卷
Guangzhou Haizhu Roundabout Tram Test Section
Station Project Research—Lingnan Picture Scroll

车站作为城市中市民参与度极高的公共交通服务设施，设计中考虑与文化结合，把车站打造成"画卷"，把岭南文化、广府文化、艺术文化等通过各种建筑材料与做法呈现于不同的车站"画卷"中，打造出极具个性的车站，成为散落在城市每一个角落的"展览馆"。

车站鸟瞰效果图

车站低点效果图

车站空间效果图

车站空间效果图

车站鸟瞰效果图

车站鸟瞰效果图

车站低点效果图

广州海珠环岛新型有轨电车试验段
车站方案研究——龙舞岭南
Guangzhou Haizhu Roundabout Tram Test Section
Station Project Research—Dragon Dance in Lingnan

车站以岭南传统民俗活动"舞龙"作为设计起点,以舞龙的动态为设计元素,创造出灵动的车站新形象,舞龙造型以下为车站空间,而舞龙造型形成了独特的屋顶廊道与平台,乘客游走其中,可在城市中得到饶具趣味的休闲体验。

车站鸟瞰效果图

车站低点效果图

车站空间效果图

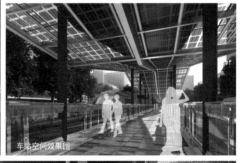
车站空间效果图

车站空间效果图

车站空间效果图

广州海珠环岛新型有轨电车试验段

车站鸟瞰效果图

广州海珠环岛新型有轨电车试验段
车站方案研究——广府趟栊
Guangzhou Haizhu Roundabout Tram Test Section
Station Project Research—Wooden Sliding Door of Cantonese Characteristic

车站以岭南民居"趟栊"门作为设计元素,提炼出"趟栊"的文化特点并融合在车站造型的设计中,通过运用绿植、光伏玻璃板、彩釉玻璃"满洲窗"等,打造一个既能传扬广府文化,又具备绿色生态特性的现代交通建筑。

车站空间透视效果图

车站低点透视效果图

车站造型构思概念分析图

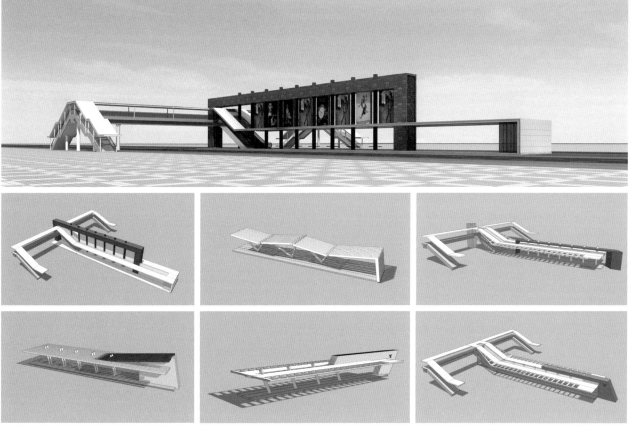

车站低点效果图

广州海珠环岛新型有轨电车试验段
车站方案研究——山岭白云
Guangzhou Haizhu Roundabout Tram Test Section
Station Project Research—Clouds on the Mountains

车站以"云与山"作为造型主要元素，顶为云，侧为山，以飘云作为车站顶棚造型，而车站立面形态结合设备空间以层叠的构件营造群山动势，同时也提炼"趟栊"理念，把山与花相互错动融合，形成独特现代的景观性交通建筑。

鸟瞰效果图

鸟瞰效果图

车站低点效果图

广州海珠环岛新型有轨电车试验段

车站鸟瞰效果图

车站低点效果图

广州海珠环岛新型有轨电车试验段
车站方案研究——广府之门一
Guangzhou Haizhu Roundabout Tram Test Section
Station Project Research—Gate of Guangzhou 1

车站立面造型表达属于广州的城市名片，立面为抽象的"广""州"。元素上通过对"趟栊"的文化传承，结合车站的使用功能与科技环保，形成一个既科技环保又简洁现代的新城市化交通建筑。

车站低点效果图

车站鸟瞰效果图

车站低点效果图

广州海珠环岛新型有轨电车试验段

车站鸟瞰效果图

广州海珠环岛新型有轨电车试验段
车站方案研究——广府之门二
Guangzhou Haizhu Roundabout Tram Test Section Station Project Research—Gate of Guangzhou 2

车站提取岭南青砖与"趟栊"元素,通过现代的设计语言穿插与变化,使现代简洁的新交通建筑重新演绎传统,在传统韵味中体现车站时代活力的特征。

车站低点效果图

车站低点效果图

车站造型构思概念分析图

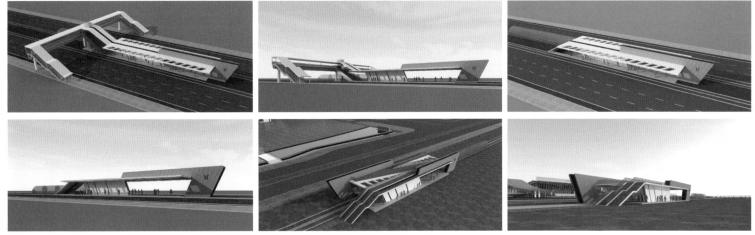

车站鸟瞰效果图

广州海珠环岛新型有轨电车试验段
车站方案研究——广府之门三
Guangzhou Haizhu Roundabout Tram Test Section
Station Project Research—Gate of Guangzhou 3

车站造型立面突出"广州"的新城市化特征，以最简洁的造型突出车站交通功能特征，结合通风装置、LED大屏、光伏玻璃顶等科技手段，使车站功能与服务更为现代与完善，为市民提供以人为本的交通服务设施。

车站低点效果图

车站低点效果图

车站鸟瞰效果图

广州海珠环岛新型有轨电车试验段
车站方案研究——云山珠水
Guangzhou Haizhu Roundabout Tram Test Section Station Project Research—Baiyun Mountain & Pearl River

车站造型通过广州风景名片"云山珠水"的理念,把车站造型打造为城市雕塑艺术装置。车站造型通过钢结构构件的组合,形成柔中带刚的车站屋顶,优美形态演绎着广州城市的独特性格。

车站低点效果图

车站低点效果图

车站鸟瞰效果图

车站鸟瞰效果图

广州海珠环岛新型有轨电车试验段
车站方案研究——城市脉搏
Guangzhou Haizhu Roundabout Tram Test Section Station Project Research—Urban Pulse

新型有轨电车作为城市交通发展的新科技产物，为城市发展提供了新力量，车站造型以"Z"形折线造型作为强有力的形态特征，寓意新型有轨电车如城市发展的脉搏，跳上城市发展的新高度。

车站低点效果图

车站鸟瞰效果图

车站造型构思概念分析图（下图）

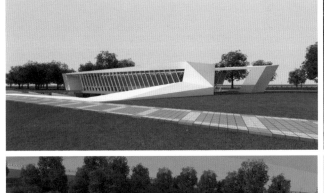

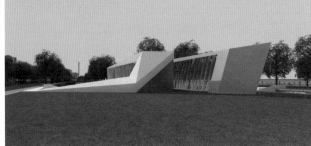

各车站造型概念设计分析（上图）

广州海珠环岛新型有轨电车试验段
车站方案研究——粤章

Guangzhou Haizhu Roundabout Tram Test Section
Station Project Research—Guangzhou Chapter

海珠环岛新型有轨电车试验段沿线尽享珠江与城市的无限风光，如奏响一首首城市的乐章。车站造型以乐章上的"彩色琴键"作为设计母题，同时与广府"趟栊门"元素结合，谱成一曲抒发广府情怀的"粤章"。

有轨电车车站优化"彩色琴键"的设计母题，车站结合绿化、艺术、广告、投影等人性化措施，通过各种不同的琴键谱写出丰富多彩的城市新乐章，形成江边与城市一道亮丽的电车风景线。

车站鸟瞰效果图

车站低点效果图

车站鸟瞰效果图

车站鸟瞰效果图

车站鸟瞰效果图

车站低点效果图

站台局部效果图

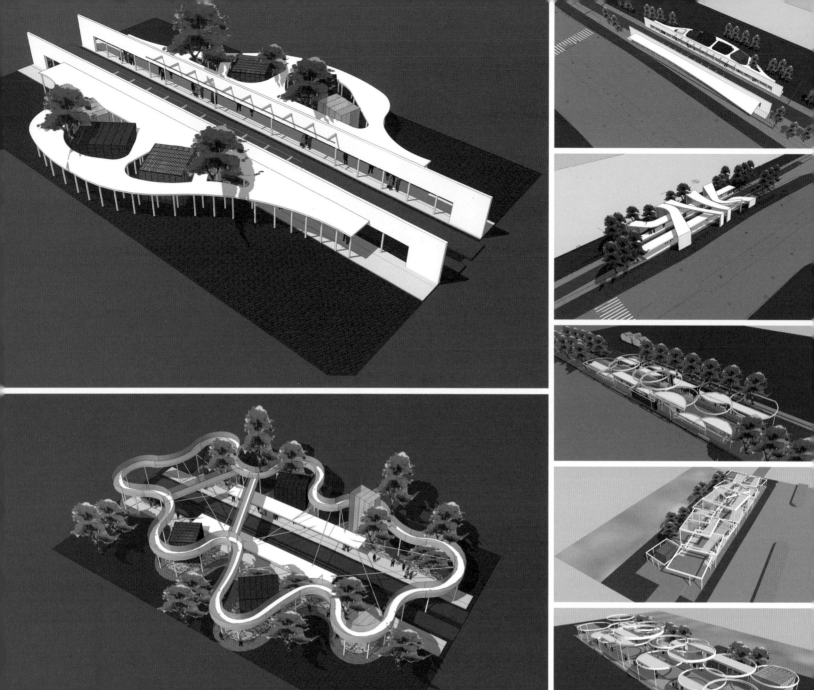

各车站概念设计分析

广州海珠环岛新型有轨电车试验段
车站方案研究——流动景点一
Guangzhou Haizhu Roundabout Tram Test Section Station Project Research—Mobile Scenic Spot 1

有轨电车车站根据不同站点的地域特征和文化主题策划，形成极具个性造型的车站。每个车站都是一个属于旅客的流动景点，市民可通过车站开展各种文化、艺术课堂等社会交流活动，形成一条由有轨电车串联而成的文化长廊。

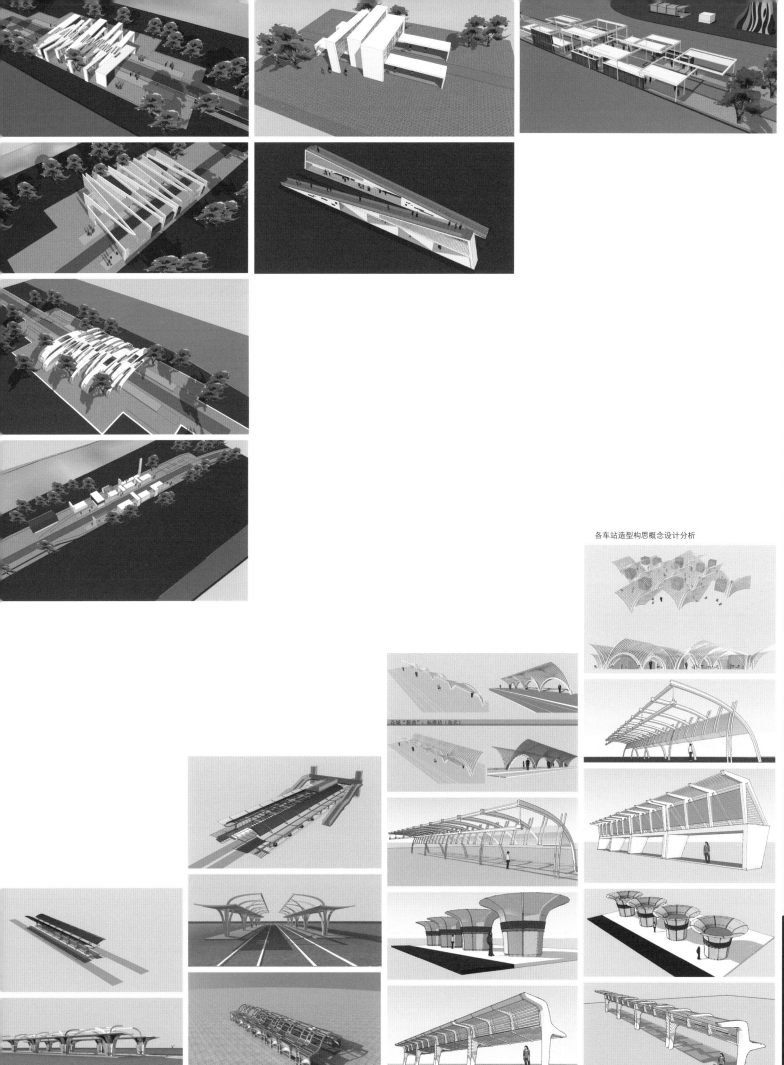

各车站造型构思概念设计分析

广州海珠环岛新型有轨电车试验段
车站方案研究——流动景点二
Guangzhou Haizhu Roundabout Tram Test Section
Station Project Research—Mobile Scenic Spot 2

在一站一景的车站方案基础上，选取三个特殊站点重点打造特色文化车站，车站通过轻盈钢结构的不同组合与构成，形成一组组城市的艺术装置，打造出属于广州与有轨电车的城市艺术结晶。

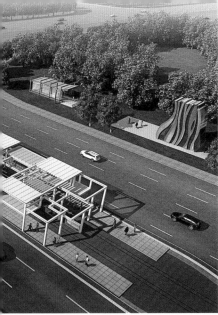

鸟瞰效果图

低点效果图

低点效果图

车站鸟瞰效果图

广州海珠环岛新型有轨电车试验段
车站方案研究——YOUNGTRAM 1
Guangzhou Haizhu Roundabout Tram Test Section
Station Project Research—YOUNGTRAM 1

广州海珠环岛有轨电车在定位上打造一条属于年轻活力的交通线路，车站以功能化、工业化、模数化、商品化作为切入点，通过轻盈的钢结构与优美的结构造型，设计出属于广州有轨电车车站的标准模式。

车站低点效果图

车站空间效果图

车站低点效果图

车站鸟瞰效果图

广州海珠环岛新型有轨电车试验段

低点效果图 | 低点效果图
鸟瞰效果图 | 鸟瞰效果图

广州海珠环岛新型有轨电车试验段
车站方案研究——YOUNGTRAM 2

Guangzhou Haizhu Roundabout Tram Test Section
Station Project Research—YOUNGTRAM 2

车站在标准模式基础上进一步深化，改进车站的结构方式与构造细节，优化车站的管线路径，从使用和技术上完善车站的实际功能，打造一套属于广州有轨电车的标准车站模式。

低点效果图

鸟瞰效果图

站空间效果图

车站空间效果图

广州滞珠环岛新型有轨电车试验段

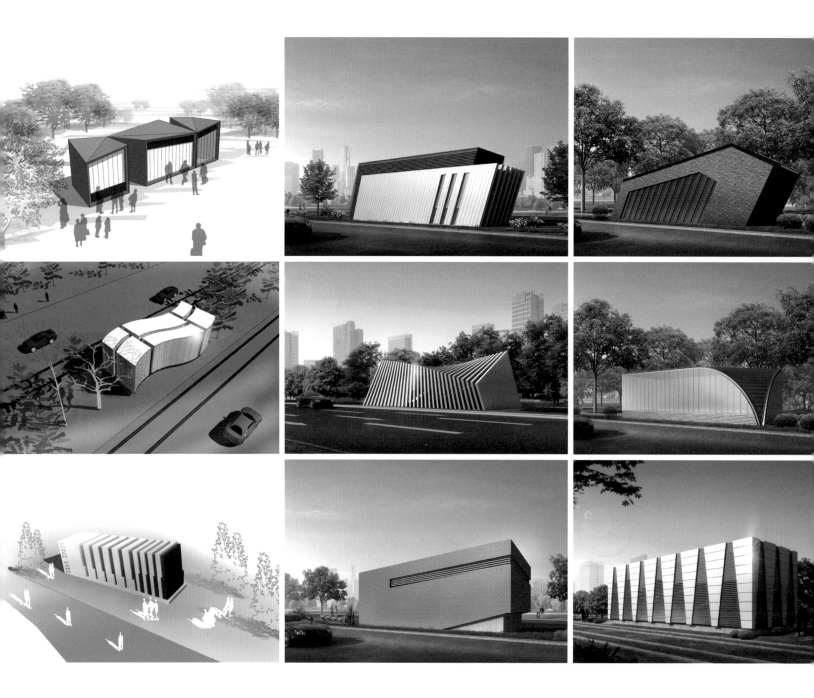

广州海珠环岛新型有轨电车试验段
有轨电车车站附属充电站方案研究
Guangzhou Haizhu Roundabout Tram Test Section
Auxiliary Charging Station Project Research

控制中心建筑景观、造型设计将传统建筑文化与现代技术、结构、材料结合起来，反映时代特点。运用简洁的建筑语言，在传统的基础上创新，在满足使用功能前提下，体现建筑自身的性格，造型设计美观大方，景观设计舒适宜人。控制中心作为地铁轨道交通配套建筑，体现了市政公共建筑的特点，也体现了地铁文化。

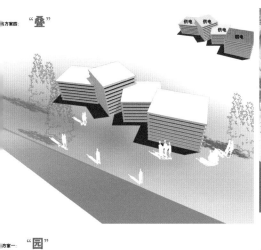

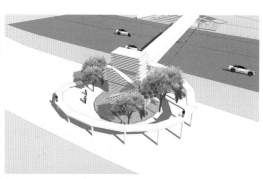

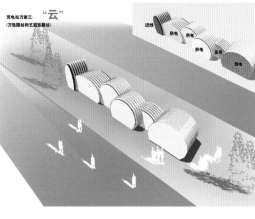

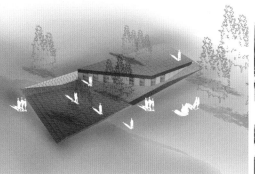
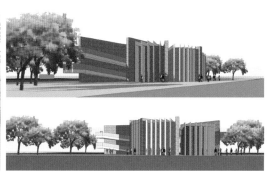
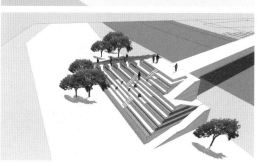

变电站总体概念

分散体量：有利于减少对城市景观的破坏

 →
原型　　　　拆散

小品化设计：提升其设计品位

 →
设备　　　　小品

改变排列方式：大大增加其体型可变性与环境适应性

根据环境组合：呼应城市环境，适应实际情况

报业中心　　　报业中心充电站方案

广州海珠环岛新型有轨电车试验段

鸟瞰效果图

低点效果图

低点效果图

广州海珠环岛新型有轨电车试验段
广州塔站充电站与折返线结合方案
Guangzhou Haizhu Roundabout Tram Test Section
Canton Tower Station Integration Plan of Charging Station and Turn-back Line

广州塔站作为线路的终点站，西侧存在折返线的大片区域，设计考虑把广州塔站的充电设施与折返线结合并开发利用，通过景观化、亲民化的方式打造交通绿化休闲一体的综合设施。首层设置充电设备空间，二层为临江公共休闲的平台空间，交通与城市相互对话，形成一道临江一线的独特风景线。

概念设计1鸟瞰效果图

概念设计2鸟瞰效果图

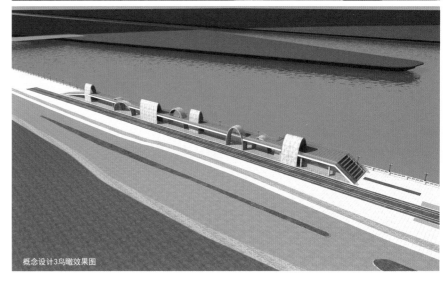
概念设计3鸟瞰效果图

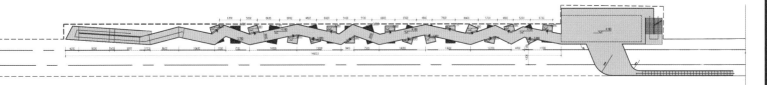
平面图

广州海珠环岛新型有轨电车试验段

低点实景

广州海珠环岛新型有轨电车试验段
有轨电车车站实施方案设计
Guangzhou Haizhu Roundabout Tram Test Section
Station Implementation Project Design

广州有轨电车试验段车站于2014年12月31日通车,为单层建筑,采用侧式站台,有效站台40m,站台宽度3m,车站不设付费区,采用站台购票、车上验票的付费方式。车站高度4.25m,最上方为车站的充电轨,满足列车充电功能。
广州新型有轨电车车站设计是从代表"羊城"的"羊角"意念出发,在满足有轨电车车站功能的前提下,通过轻盈的钢结构单悬挑形式表达简洁现代的有轨电车车站造型,柔美灵动的车站造型与开敞通透的车站站台辉映。车站建筑之轻和透是应对岭南气候的重要设计方式,也是一种地域性的理性表达。

车站空间实景

低点实景

低点实景

广州海珠环岛新型有轨电车试验段

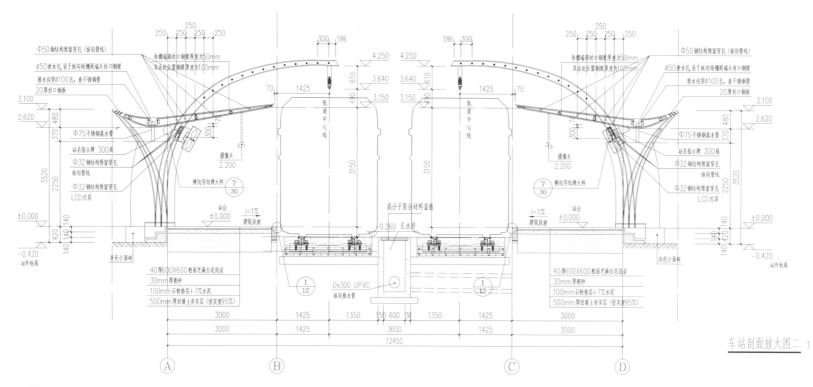

剖面图

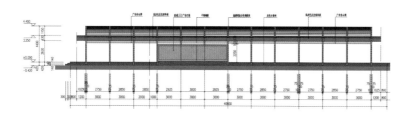

立面图

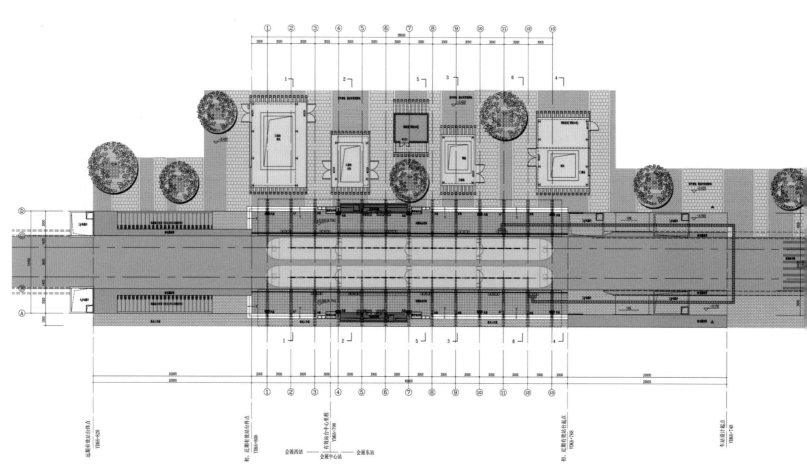

平面图

附属建筑实景

附属建筑实景

附属建筑实景

建筑结构局部实景

建筑结构局部实景

沿街低点实景

有轨电车建筑泛光照明设计1

有轨电车建筑泛光照明设计2

有轨电车建筑泛光照明设计3

有轨电车建筑泛光照明设计4

广州海珠环岛新型有轨电车试验段
有轨电车建筑泛光照明设计
Guangzhou Haizhu Roundabout Tram Test Section
Building Floodlighting Design

在有轨电车车站及区间的泛光照明设计上，遵循"动+静"的设计理念。车站建筑属于列车和客流停留的空间，泛光通过多层次反射照明，塑造车站空间静态氛围；区间属于列车快速通过区域，泛光简洁明快处理，强调快速通过所形成的动态效果。动静间的切换符合交通逻辑与功能逻辑，完成交通属性的心理暗示。

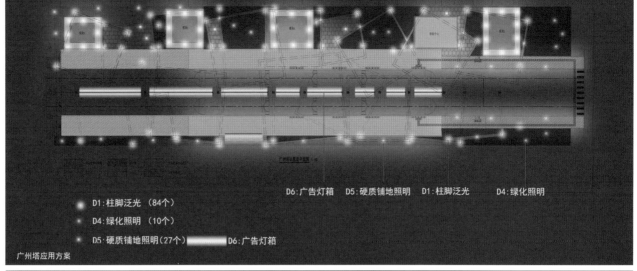

D1：柱脚泛光（84个）
D4：绿化照明（10个）
D5：硬质铺地照明（27个） D6：广告灯箱

广州塔应用方案

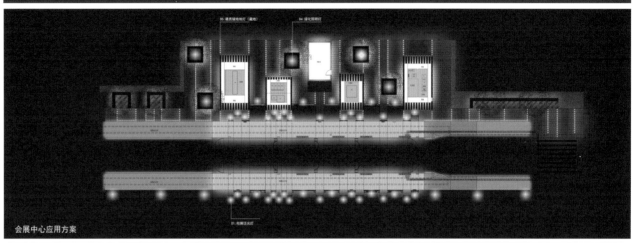

会展中心应用方案

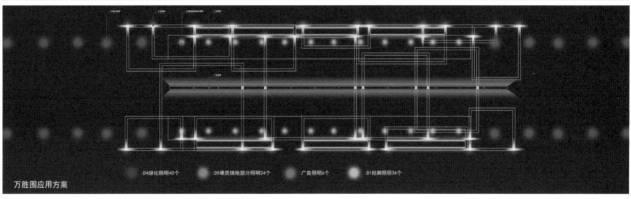

万胜围应用方案

折返线应用方案

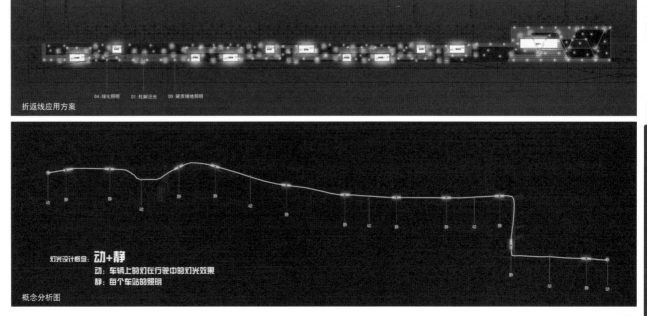

灯光设计概念：**动+静**
动：车辆上的灯在行驶中的灯光效果
静：每个车站的照明

概念分析图

环境实景1

广州海珠环岛新型有轨电车试验段
琶醍站改造设计
Guangzhou Haizhu Roundabout Tram Test Section
Pati Station Renovation Design

琶醍站配套附属建筑包括旧码头上盖平台与琶醍中区商铺两个部分。其中，旧码头上盖平台配套工程为区间上方2层的景观平台，利用原啤酒厂室外码头结构，在既有结构上架设2层的钢结构平台。平台首层架空，为有轨电车的区间轨行区，二层平台为一线江景平台，将首层轨行区占据的江边休闲空间还给二层。项目在线路设计与现场条件冲突的情况下，通过改造更新，满足最终功能要求，留下既有环境记忆，减少破坏，同时也让园区内的新元素——有轨电车站及区间轨行区，更好地融入周边环境中。琶醍中区商铺是电车项目拆迁的复建商铺，采用钢结构形式，单层建筑，局部3层，建筑立面通过玻璃幕墙、铝板及铝格栅等现代材料及简洁的形式，融合到周边的艺术园区内，并结合二层室外屋顶平台，为周边提供一个江边的室外休闲空间。

环境实景2

实景1

实景2

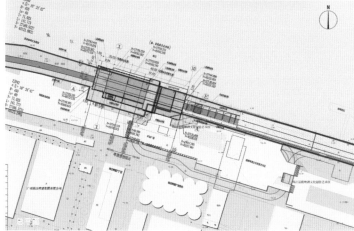
总平面

实景4

出入口实景

广州海珠环岛新型有轨电车试验段
琶醍站配套附属建筑设计
Guangzhou Haizhu Roundabout Tram Test Section Pati Station Auxiliary Building Design

琶醍站为琶醍艺术创意园区内的原珠江啤酒厂水泵房。水泵房原状为一座10m高的单层建筑物，地上1层，地下1层，地下局部设置了夹层。车站通过改造，重新利用原水泵房主体结构。在首层板通过加固，组织车站轨行区及站台空间。5.5m标高上通过增加钢结构楼板，形成两层空间，作为车站配套功能，并在原结构屋面上通过加固，局部增设三层空间。琶醍车站通过创新性的改造处理，形成一个通过工业建筑改造而成的车站交通建筑。并通过工业化的立面元素，现代的设计手法，还原建筑原有的工业气息。

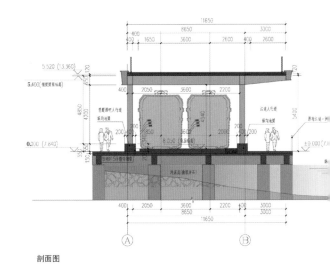

剖面图

站外堤岸环境实景

轨道空间实景

轨道与人行交汇点实景

站外环境实景

站止休闲环境实景

广州海珠环岛新型有轨电车试验段

鸟瞰效果图

广州海珠环岛新型有轨电车试验段
琶醍区域立体综合交通枢纽方案研究
Guangzhou Haizhu Roundabout Tram Test Section
Pati Area Multimodal Transport Hub Project Research

广州海珠环岛有轨电车试验段除经过广州塔地标外，经过的另外一个重要片区就是琶洲电商总部与琶醍啤酒文化特色餐饮区。结合有轨电车项目，通过"城市绿T"的规划概念，为片区打造一个T型的全新休闲文化平台。而"城市绿T"的核心项目是琶醍文创区。琶醍文创区作为基于珠江啤酒厂与琶醍啤酒区的整体式文化创意工业改造片区。是一个具有时代示范性与影响力的创新型项目。项目通过休闲平台与创意建筑及城市空间整合了工业文化、交通文化、创意文化、艺术文化、绿色文化的各种可能性，文化在此处的大融合与各方的活动交流如化学反应一样产生无限的可能，大大提升了片区的城市活力，形成了城市独有的文化标杆。

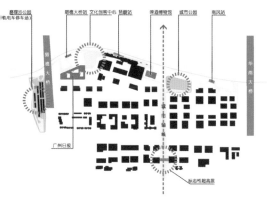

T LANE

案分析图

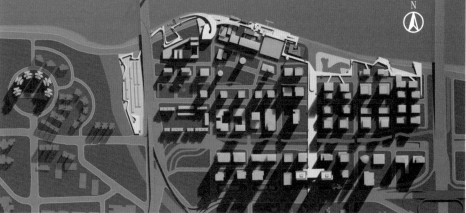
总平面图

广州海珠环岛新型有轨电车试验段

鸟瞰效果图

局部低点效果图

鸟瞰效果图

概念分析图

"潮"

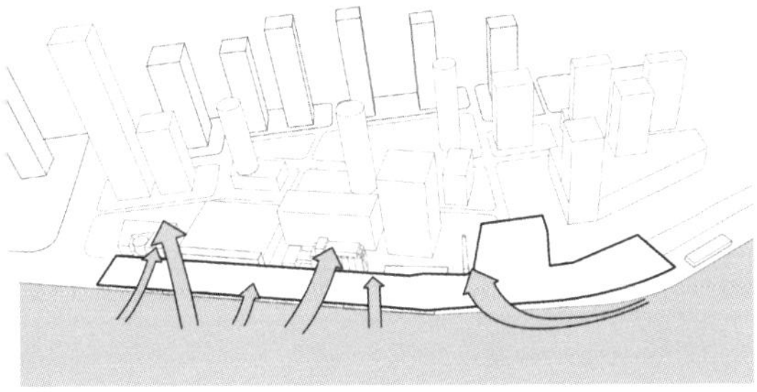

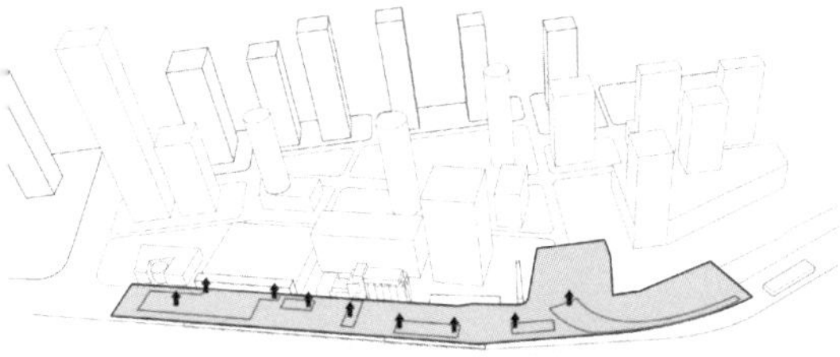

鸟瞰效果图

局部低点效果图

鸟瞰效果图

局部低点效果图

鸟瞰效果图

鸟瞰效果图

局部低点效果图

局部低点效果图

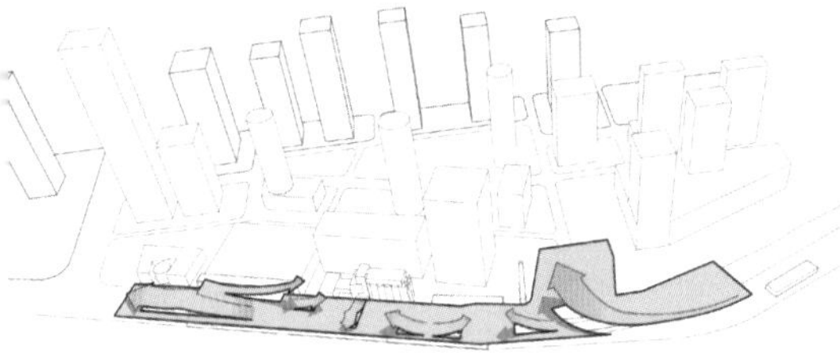

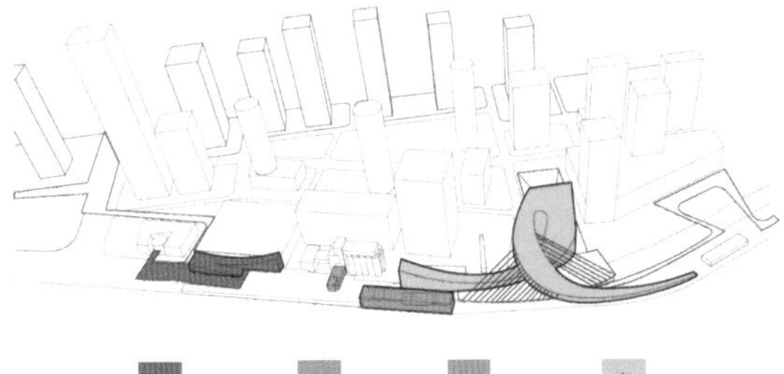

台（创意中心）　亭（传媒中心）　舫（美术中心）　廊（演艺中心）

鸟瞰效果图

鸟瞰效果图

部低点效果图

低点效果图

投标设计阶段概念分析图

酝酿-激活因子

方　　　　坊　　　　FUN

鸟瞰效果图

低点效果图

总平面图

鸟瞰效果图

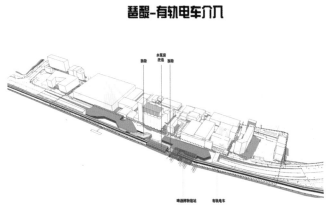

区域改造分析图

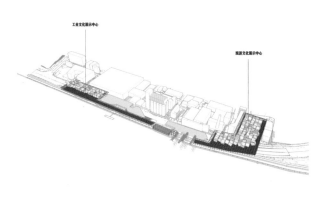

绿道

5m 标高人流线

局部效果图

局部效果图

局部效果图

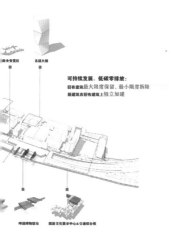

可持续发展，低碳零排放：
旧有建筑最大限度保留、最小限度拆除
新建筑在旧有建筑上独立加建

绿道

5m 标高人流线

水上码头换乘流线

流线分析图

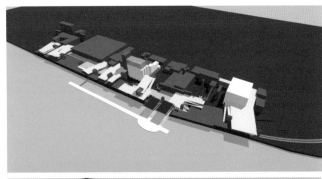

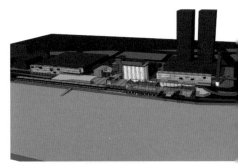
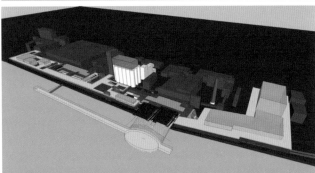

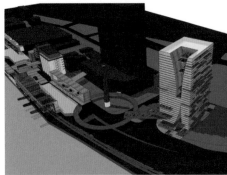
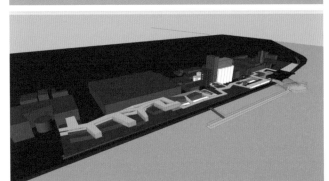

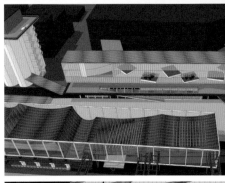
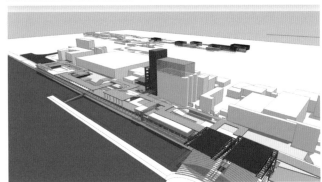
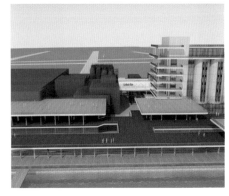

设计分析图

简单设计

简单，反映的是一种生活态度，同样也是一种设计的态度。
"简单设计"不是"简陋设计"，一个"简单设计"是设计者对现实归纳与提炼的智慧成果。
作为建筑设计者，我们一直试图寻找一种属于建筑的"简单设计"。
广州市有轨电车文化体验中心（项目暂定名），是一个地处珠江边上的文化交通建筑。受业主委托，我们是设计方。由于地处珠江啤酒厂西侧，广州报业大厦东侧的珠江边上，地理位置得天独厚（一线江景），而且业主非常关注建筑体验感，因此建筑设计对于设计者来说是一个巨大的挑战。
尝试着用"简单设计"思考项目的各种可能性。简单体块，外加单纯设计逻辑，形成一个讲求内外空间变化，富有故事性的建筑物，并自然产生各种充满趣味的体验空间。我们提炼出 10 个方案，用一个字来代表方案的设计核心，分别是"谷、叠、柱、游、影、透、旋、开、门、搭"。

谷 透
叠 旋
柱 开
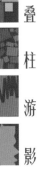游 门
影 搭

造型设计方案分析图

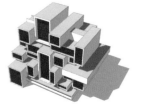
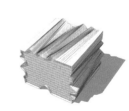
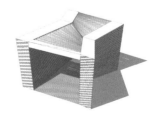

鸟瞰效果图

低点效果图

广州海珠环岛新型有轨电车试验段
有轨电车车辆段方案研究
Guangzhou Haizhou Roundabout Tram Test Section
Tram Deport Project Research

有轨电车车辆段停车场位于广州海珠区磨碟沙公园范围内，周边环境优越，停车场的规划提出以"电车公园"作为规划概念，将停车场与公园融为一体，作为开放给市民的科普基地，摆脱了传统停车场封闭、简易等弊端。规划将停车场该有的建筑体量打散，分散布置于公园内，将停车场的功能建筑打造成公园内的参观景点，成为相互依存的环境关系。

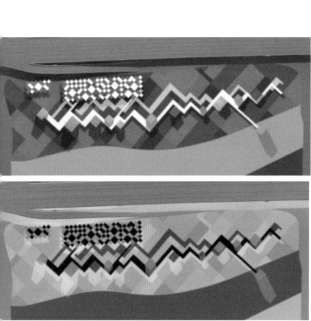
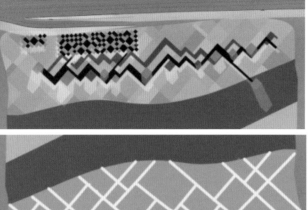

鸟瞰效果图

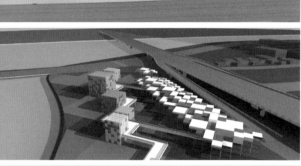
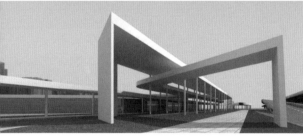

鸟瞰效果图

低点效果图

广州海珠环岛新型有轨电车试验段

广州海珠环岛新型有轨电车附属配套
公共卫生间
Guangzhou Haizhu Roundabout Tram
Ancillary Facility
Public Washroom

有轨电车线路经过江边绿化带区域，需对原绿化带中附属配套公共卫生间进行迁改与复原，在设计中，考虑将配套设施景观化处理，通过圆形体量化解建筑与景观之间的矛盾，公共卫生间与环境之间通过灰空间连接过渡，使其更好地融入绿化带环境之中。

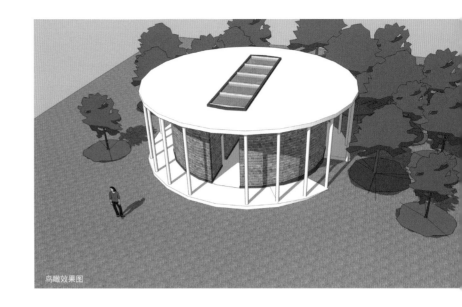

鸟瞰效果图

广州海珠环岛新型有轨电车附属配套
琶醍中心区改造
Guangzhou Haizhu Roundabout Tram
Ancillary Facility
Central Pati Renovation Design

有轨电车线路经过琶醍区域，线路区间对原有琶醍酒吧商铺局部产生影响，设计对原商业布局重新规划，在满足商铺原面积回迁的同时，创造出工业主题更为浓厚的单层商业建筑，利用建筑顶部形成江景视野更为开阔的两层平台，为游客提供最优质的江景体验。

改造实景

广州海珠环岛新型有轨电车附属配套
天桥方案设计
Guangzhou Haizhu Roundabout Tram
Ancillary Facility
Overpass Project Design

广州塔区域和有轨电车广州塔站需增设一组过路天桥以解决此区域的交通衔接问题，在天桥设计上，考虑把广州塔、基座平台、有轨电车站、地铁出入口等元素一并统筹考虑，在解决交通功能同时兼顾景观与休闲，在广州塔地标下创造出舒适宜人的交通设施。

鸟瞰效果图

节点效果图

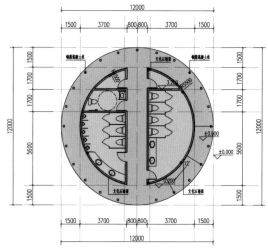
平面图

改造效果图

改造实景

节点效果图

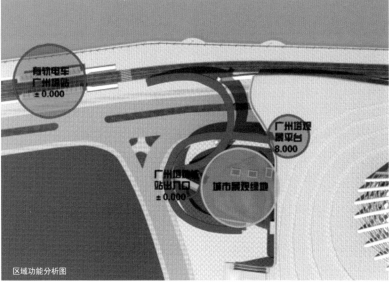
区域功能分析图

优化概念鸟瞰效果图

广州海珠环岛新型有轨电车试验段
广州塔周边结合车站配套设施环境优化设计
Guangzhou Haizhu Roundabout Tram Test Section
Environmental Optimization Design of Canton Tower Peripheral Area Combing with Station Ancillary Facility

控制中心建筑景观、造型设计，将传统建筑文化与现代技术、结构、材料结合起来，反映时代特点。运用简洁的建筑语言，在传统的基础上创新，在满足使用功能的前提下，体现建筑自身的性格。造型设计美观大方，景观设计舒适宜人。控制中心作为地铁轨道交通配套建筑，体现了市政公共建筑的特点，也体现了地铁文化。

优化概念低点效果图

优化概念低点效果图

"广"纳百川

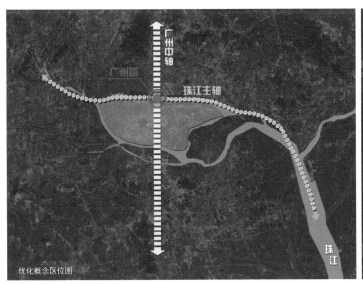

优化概念区位图

优化概念总平面图

优化概念低点效果图

优化概念鸟瞰效果图

优化概念分析图

优化概念低点效果图

优化概念低点效果图

交通综合体

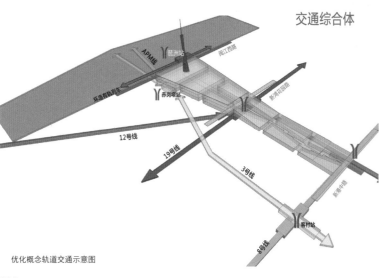
优化概念轨道交通示意图

优化概念鸟瞰效果图

后记 Postscript

从最初计划到审编已历时一年有余。
所谓隔行如隔山。原来出一本书并不容易。
资料收集、整理、拍照、排版、文字、磨合、修改、讨论，甚至争吵，每一步都比设想的要困难。
最初计划及安排也不得不一变再变。
哲人叔本华有名句"要么庸俗，要么孤独"。从开始有出书的设想，已经注定要走上一条孤独的道路，做前人未做过、自己要尝试的事。
好在有心中坚守的信念和目标，以及志同道合的伙伴一路前行。
整理资料，看看过去经过的每一个案例，仍会唤起尘封的记忆，鼓励着大家，克服惰性和困难，以责任和担当，展现对轨道交通事业的激情。
回望来时的路，已逝者如斯夫。
谨以此书，感谢设计团队及伙伴们的努力与付出！
感谢院领导对设计工作的一贯重视及支持！
感谢业主一直不变的厚爱与信任！
感谢所有合作单位一路上的风雨同行，
感恩新时代为我们创造了无限的机遇与未来。

It has been over one year from initial planning to editing of this book. Difference in profession makes one feel worlds apart, and it was never easy to get a book published.

Collecting information, organizing materials, taking pictures, designing layouts, writing content, collaborating, amending, discussing and even arguing, every step is harder than imaged.

And the planning and arrangement has been changed again and again. There is a quote from the philosopher Arthur Schopenhauer, "Being Mediocre, or being lonely". Ever since the idea of publishing a book came into being, I was destined to be lonely, to do something I want to try on my own that nobody had done.

Fortunately, the faith and goal was kept so that I could proceed with my like-minded friends.

Memory was revoked when organizing the materials and reviewing each past case, which kept encouraging us to overcome laziness and difficulties, to express our passion on rail transit career with responsibilities and commitments.

Thus things flow away day and night when I look back on the way. This book is dedicated to the efforts and contributions of the design team and workmates.

I would also like to thank the leaders of the Architectural Design and Research Institute of Guangdong Province, for their attention and support for our design.

Thank the proprietors for their great kindness and confidence as always.

Thank all cooperating organizations for journeying together against all odds.

Thank the new era for giving us infinite opportunities and future.

团队 Team

图书在版编目（CIP）数据

轨·道：交通建筑作品 / 曾宪川等主编. —北京：中国建筑工业出版社，2019.4
ISBN 978-7-112-23498-1

Ⅰ.①轨… Ⅱ.①曾… Ⅲ.①轨道交通—交通运输建筑—建筑设计—作品集—中国—现代 Ⅳ.①TU248

中国版本图书馆CIP数据核字（2019）第049868号

责任编辑：费海玲　张幼平
特邀策划：叶　飚
翻　　译：黄健茵
版面设计：叶仲轩　麦韵心　陈　峻　刘　澳
责任校对：王　烨

主编：曾宪川　陈　雄　江　刚　罗若铭
编委：陈冠东　钟仕斌等
汇编：广东省建筑设计研究院
策划：广州维图轩广告设计有限公司

轨·道　交通建筑作品
主编　曾宪川　陈　雄　江　刚　罗若铭
广东省建筑设计研究院
*
中国建筑工业出版社出版、发行（北京海淀三里河路9号）
各地新华书店、建筑书店经销
广州维图轩广告设计有限公司制版
恒美印务（广州）有限公司印刷
*
开本：965×1270毫米　1/16　印张：25　字数：1238千字
2019年11月第一版　2019年11月第一次印刷
定价：360.00元
ISBN 978-7-112-23498-1
　　　（33793）

版权所有　翻印必究
如有印装质量问题，可寄本社退换
（邮政编码100037）